Mercury

Mercury

Rising Tales Of Temperature

Rayan Musk

Mohammed Altaf Hussain

CONTENTS

INDEX

Chapter 5: Celsius Conspiracy

5.1 Uncover a sinister plot that involves manipulating temperature for personal gain or control.

5.2 The team embarks on a thrilling adventure to unravel the conspiracy.

5.3 Face internal and external challenges that test the bonds within the group.

Chapter 6: Kelvin's Redemption

6.1 Dive into the backstory of a seemingly antagonistic character, revealing their motivations and struggles.

6.2 Develop empathy for the antagonist, adding depth to the overall narrative.

6.3 The protagonist faces moral dilemmas and questions the true nature of their power.

Chapter 7: Absolute Zero Hour

7.1 Reach a climactic point where the group confronts the mastermind behind the temperature conspiracy.

7.2 Uncover the true potential and limitations of the protagonist's power.

7.3 Explore themes of sacrifice, loyalty, and the consequences of wielding such a unique ability.

Chapter 8: Equilibrium Epiphany

8.1 After the confrontation, the group reflects on their journey and the lessons learned.

8.2 The protagonist comes to terms with their identity and the responsibility that comes with their power.

8.3 Tie up loose ends and prepare for the resolution of the overarching plot.

Chapter 9: The Eternal Flame

9.1 Conclude the story with a resolution to the larger conflict and the fate of the world.

9.2 Reflect on the personal growth of the protagonist and the impact of their choices.

9.3 Leave room for potential sequels or spin-offs, hinting at new adventures in the world of temperature manipulation.

Conclusion

Introduction

In the heavenly expressive dance of our planetary group, one humble sphere charms our creative mind with its double nature - a rankling vicinity to the sun and a freezing void that stretches into the vast pit. Mercury, the littlest and nearest planet to the sun, moves on the edge of limits, typifying the story of temperature in a grandiose story that winds around logical interest with the idyllic charm of the unexplored world.

From the get go, Mercury introduces itself as an infertile and cold world, a burned scope where the sun overwhelms the sky with unwavering savagery. Its surface, scarred by cavities and gorges, recounts an account of persistent sunlight based barrage. The sun, a blasting god in Mercury's vast theater, releases a deluge of radiation that prepares the planet's rough landscape to temperatures that challenge natural understanding. Here, under the constant look of our heavenly heater, the mercury in thermometers would shrink, unfit to comprehend the stunning intensity that overruns Mercury's days.

However, as the sun plunges beneath the skyline, a chilling conundrum unfurls. Mercury, without a thick environment to hold heat, dives into a nighttime profound freeze. In this astronomical swing between limits, the planet's surface temperature decreases steeply, plummeting to levels that would freeze water, however the actual embodiment of plausibility. The story of Mercury's temperature is an arresting adventure, a heavenly yin and yang where limits impact, making a powerful ensemble that reverberates with the reverberations of logical request and human interest.

This transaction of intensity and cold on Mercury welcomes us to test the secrets of our planetary group's deepest planets. Past the shallow dance of heavenly bodies, it entices us to unwind the complex powers that shape the predetermination of a world suspended in the grandiose expressive dance. The temperature story of Mercury, carved across its surface like an enormous unique finger impression, is a gateway into the more profound openings of planetary science, where the laws of physical science and the impulses of divine mechanics combine in an entrancing transaction.

To comprehend the mercury ascending on this baffling world, we set out on an excursion through the hallways of logical investigation. The story unfurls not just as a narrative of temperature changes but rather as an investigation of the human soul that drives us into the unexplored world. It is a story of strength and disclosure, where the intensity of request addresses the frosty difficulties of the universe. As we dig into Mercury's secrets, we wind up crossing the domains of astronomy, geography, and barometrical science, each discipline adding to the rich embroidered artwork of figuring out that winds around the texture of this inestimable account.

The sun, our heavenly anchor, assumes the part of hero and bad guy in Mercury's temperature adventure. Its brilliant energy, the backbone of our nearby planet group, energizes the extraordinary intensity that characterizes Mercury's days. The vicinity of this little planet to the sun is both a gift and a revile, a sensitive harmony between nurturing warmth and searing force. As we stand on the cliff of Mercury's sunlit domain, we witness the heavenly artful dance that unfurls between the star and its planetary sidekick.

Mercury's days are an orchestra of sun powered radiation, an unrelenting surge that washes the planet's surface in an outpouring of energy. The sun, a burning sphere in the Mercurian sky, barrages the rough scene with photons that convey the pith of intensity. In contrast to Earth, where a defensive climate diffuses and controls sun powered radiation, Mercury stands uncovered, its surface retaining the full brunt of the sun's rage.

Subsequently, daytime temperatures on Mercury take off to inconceivable levels, arriving at up to a burning 800 degrees Fahrenheit (427 degrees Celsius). This diabolical scene, where the very ground underneath one's feet takes steps to dissolve, challenges our previously established inclinations of planetary circumstances. It is where lead would melt, and conventional thermometers would give up despite such burning intensity.

The shortfall of a significant climate on Mercury worsens the temperature limits. In contrast to Earth, where our environment goes about as a warm cover, catching and rearranging heat, Mercury's dainty cloak of gases gives little rest. The planet's weak air, made predominantly out of hints of helium, hydrogen, and oxygen, is no counterpart for the tenacious sunlight based surge. Therefore, Mercury encounters a glaring difference between its sunlit days and shadow-hung evenings.

The change from day to night on Mercury is an infinite artful dance of quick temperature variances. As the sun plunges beneath the skyline, the planet goes through a quick transformation, shedding the intensity of the day into the cold profundities of room. Without any a climate to hold and course warmth, Mercury's surface cools quickly, plunging evening temperatures to chilling profundities. The mercury in thermometers, in the event that it could endure the rankling day, would contract notwithstanding this nighttime cold, illustrating the planet's dualistic nature.

The Mercurian night is a cold domain where temperatures can plunge to a frigid less 290 degrees Fahrenheit (short 179 degrees Celsius). In this cryogenic scene, where the very air stands frozen, the actual ground turns into a ruined spread of frigid destruction. The temperature swing from day to night on Mercury is definitely not a continuous drop however a sharp dive, a demonstration of the shortfall of directing powers that describe our more calm heavenly house.

As we explore the limits of Mercury's temperature woven artwork, we end up facing the perplexing idea of its surface elements. The planet's scene, scarred by ages of infinite barrage, recounts an account of topographical tumult and brutal disturbance. The transaction of intensity and cold on Mercury's surface is carved in the very shakes and pits that demonstrate the veracity of the planet's turbulent history.

One of the characterizing elements of Mercury's scene is its sweeping fields, immense stretches of moderately smooth territory that give a false representation of the fierce powers that formed them. These fields, known as "smooth fields," are remembered to result from volcanic movement that reemerged enormous bits of the planet. The interaction among heat and liquid stone on Mercury's surface offers a brief look into the unique powers that shape planetary scenes.

Confounding the fields are gigantic scarps, transcending precipices that resist the customary way of thinking of planetary geography. These precipices, known as "lobate scarps," are believed to be the consequence of the planet's compression as it cooled and hardened. The limits of temperature on Mercury, from burning intensity to frigid cold, have influenced the planet's geology, forming highlights that challenge how we might interpret planetary elements.

As we test further into Mercury's land history, we experience another puzzler - the presence of water ice in the planet's polar areas. In a scene where temperatures swing between limits, the presence of frozen water appears to be confusing. Mercury, with nearness to the sun and daytime temperatures could liquefy lead, seems a far-fetched have for cold stores.

However, the pits close to Mercury's posts cast an alternate story. In these ceaseless shadows, where the sun's beams never reach, temperatures stay low enough for water ice to endure. The disclosure of water ice on Mercury opens another part in our investigation of the planetary group, bringing up issues about the beginning and movement of unstable substances in our astronomical area. It challenges our presumptions about the conveyance of water in the deepest locales of our planetary group, where the sun's blazing hug appears to be an improbable partner to frozen compounds.

The dance of temperature on Mercury isn't bound to its surface alone; it reaches out into the domains of its shaky climate. While scanty and wispy, Mercury's exosphere, the peripheral layer of its air, assumes a pivotal part in the planet's temperature elements. The interaction between the sun and this delicate envelope of gases adds one more layer to the inestimable story, a fragile dance of particles and photons that characterizes the warm expressive dance on Mercury.

The sun based breeze, a flood of charged particles shot out by the sun, besieges Mercury's questionable environment with steady power. In contrast to Earth, where a hearty attractive field safeguards us from the sun oriented breeze, Mercury stands presented to its full power.

The crash between sun powered breeze particles and the gases in Mercury's exosphere establishes a unique climate where intensity is traded and temperatures vacillate.

Notwithstanding the sun powered breeze attack, Mercury's exosphere answers with astounding flexibility. In spite of its slimness, this wispy layer of gases figures out how to hold a piece of the sun's energy. The exchange between sun powered radiation and the exospheric gases prompts temperature varieties, establishing a unique warm climate that reaches out past the planet's surface.

The presence of unstable mixtures in Mercury's exosphere adds one more layer to the temperature story. The planet's exosphere contains hints of components like sodium, potassium, and calcium, which can assimilate and once again transmit sun powered energy. This communication between unpredictable components and sun based radiation adds to the intricate temperature elements that shape Mercury's environment, a sensitive expressive dance of intensity ingestion and emanation that resists the planet's apparently ungracious nature.

As we unwind the complexities of Mercury's temperature account, we wind up at the junction of logical request and human creative mind. The limits of intensity and cool, the powerful interaction of geographical powers, and the baffling presence of water ice challenge how we might interpret planetary science. The tale of Mercury's temperature isn't just a narrative of actual peculiarities; it is an investigation of the limits of human information and the endless mission to disentangle the secrets of the universe.

In this enormous odyssey, we go up against the restrictions of our mechanical ability as we send space apparatus to circle and concentrate on Mercury very close. Missions like NASA's Courier (Mercury Surface, Space Climate, Geochemistry, and Going) have given extraordinary experiences into the planet's structure, geography, and environmental elements. These mechanical messengers, furnished with state of the art instruments, have turned into our eyes and ears in the cruel environs of Mercury, conveying information that energizes the flares of logical interest.

The innovative wonders that empower our investigation of Mercury likewise epitomize the dauntless human soul. The difficulties of sending space apparatus to the deepest ranges of our nearby planet group, where the sun's gravitational hold is generally serious, test the constraints of designing and inventiveness. From the singing intensity of send off to the cold profundities of interplanetary space, these missions push the limits of what is conceivable, reflecting the limits that characterize Mercury's own vast adventure.

As we dig further into the story of Mercury's temperature, we stand up to the more extensive ramifications for how we might interpret planetary frameworks past our own. The illustrations gained from Mercury's limits resound with our journey to disentangle the secrets of exoplanets, far off universes that circle stars past our nearby planet group.

The exchange of intensity and chilly, the elements of airs, and the geographical powers that shape planetary scenes find reverberations in the most distant scopes of the universe.

In the consistently growing embroidery of galactic information, Mercury's temperature story fills in as a standard, a perspective that secures our investigation of the universe. It welcomes us to contemplate the variety of planetary circumstances and the heap manners by which the powers of the universe shape the universes we experience. The limits of temperature on Mercury, a long way from being a peculiarity, become an impression of the grandiose dance that oversees the predeterminations of heavenly bodies all through the immense region of room.

As we explore the exciting bends in the road of Mercury's temperature story, we wind up remaining at the junction of logical disclosure and human creative mind. The story unfurls not similarly as a narrative of temperature variances on a far off world yet as a demonstration of the human soul that pushes us into the unexplored world. Mercury, with its limits of intensity and cold, coaxes us to push the limits of our comprehension, to wander past the agreeable bounds of our earthbound house and investigate the vast miracles that anticipate.

Eventually, Mercury rising recounts to a story that rises above the limits of our nearby planet group, venturing into the profundities of our aggregate interest and the boundless spread of human potential. It is a story of temperature, indeed, yet additionally a story of industriousness, strength, and the steady quest for information that characterizes our excursion through the universe. As we look at Mercury, a little world suspended in the endlessness of room, we track down motivation to proceed with our mission for understanding, to disentangle the secrets that lie into the great beyond, and to embrace the difficulties that look for us on this vast excursion.

Chapter 1

The Frozen Prelude

In a domain where time appeared to have surrendered its unending walk, where the air balanced weighty with the glasslike embrace of winter, there existed a spot known as Eldrath. It was a land hidden in an everlasting ice, a demonstration of the eccentric idea of the components that represented its fate. Eldrath's scene bore the scars of endless snowstorms, every snowflake a quiet observer to the entry of ages.

At the core of Eldrath stood the Frosty Bastion, a glorious post that rose from the frozen ground like a landmark to the unstoppable soul of its occupants. The bastion's transcending towers came to towards the sky, their frosty pinnacles penetrating the sky blue sky. Inside its walls, a general public of versatile spirits, known as the Cryomancers, had cut out a development in the midst of the unforgiving virus.

The Cryomancers, bosses of ice and snow, had figured out how to saddle the cold powers that wrapped their reality. It was a fragile hit the dance floor with the components, a cooperative relationship that conceded them both power and insurance. In Eldrath, the seasons held no influence, for winter ruled, and the Cryomancers were its stewards.

As the occupants of Eldrath approached their regular routines, there was a propensity of pressure all around. Murmurs of an old prescience had started to flow among the Cryomancers — a prediction that anticipated of an extraordinary commotion, a disastrous occasion that would shake the underpinnings of their frosty realm.

The Frozen Introduction, as the prescience was known, discussed a particularly favored one who might rise out of the profundities of Eldrath's chilling breezes. This favored one was bound to use a power not at all like whatever other, a power that could either save their reality from looming destruction or dive it into a never-ending winter.

Eldrath's ruler, Cryomancer Preeminent Selene Frostheart, considered the expressions of the prescience in the isolation of her ice-cut chamber. Her silver hair flowed like a frozen cascade, and her eyes, the shade of frosty ice, mirrored the heaviness of the predicting. Selene, shrewd and respected, realize that the Frozen Introduction couldn't be disregarded.

In the core of Eldrath, an unassuming town settled between ice loaded trees, a youthful Cryomancer named Aelar felt the call of predetermination reverberating in the profundities of his spirit. Aelar, with his raven-dark hair and puncturing blue eyes, had a natural association with the essential powers that encompassed him. Unbeknownst to him, he was the anointed one talked about in the prediction.

Aelar's process started when an otherworldly figure, washed in an ethereal shine, appeared to him in a fantasy. This unearthly courier uncovered the prescience's privileged insights, unfurling the strings of destiny that bound Aelar to the predetermination of Eldrath. The youthful Cryomancer stirred with a newly discovered reason, his heart burning sincerely.

Directed by the murmurs of the prescience, Aelar set out on a mission to unwind the secrets of Eldrath's old past. He crossed the frozen breadths, experiencing both the amazing excellence and hazardous difficulties that lay on pause. The scene, when natural, turned into a steadily moving maze of ice and shadows.

As Aelar wandered further into Eldrath, he found failed to remember sanctuaries covered underneath layers of ice, their walls embellished with perplexing glyphs that discussed when the Cryomancers' authority over the components was unchallenged. These disclosures energized's comprehension Aelar might interpret his own capacities, opening torpid powers that flooded through his veins like fluid ice.

However, with each bit nearer to his predetermination, Aelar confronted foes brought into the world from the actual pith of Eldrath's baffling sorcery. Ice golems, energized by old spells, ambled forward to test the youthful Cryomancer's fortitude. Phantoms of ice, ghostly leftovers of past Cryomancers, tried to obstruct Aelar's advancement with ethereal ringlets of chilling energy.

In the midst of these preliminaries, Aelar experienced partners who perceived the meaning of his excursion. Cryomancers of various ages and encounters went along with him, their aggregate assets framing a mosaic of force that reflected the variety of Eldrath itself. Together, they confronted the difficulties that lay ahead, fashioning bonds as tough as the ice that encompassed them.

As the gathering dug further into Eldrath's insider facts, they revealed an old relic — the Frostheart Gem. This translucent artifact, throbbing with a supernatural gleam, held the way to opening Aelar's actual potential. The Cryomancers understood that the precious stone was a point of convergence of the Frozen Introduction, an impetus for the looming predetermination that looked for them all.

Be that as it may, the Frostheart Gem was not effortlessly gotten. Protected by a legendary watchman, a titanic animal of ice and ice known as Glacius the Frostbound, the gem lay inside the core of the Cold Fortress. Glacius, a sentinel of Eldrath's past, had rested in everlasting cautiousness until the time predicted in the Frozen Preface showed up.

The showdown with Glacius tried the Cryomancers' purpose. The animal, energized by the exceptionally wizardry that saturated Eldrath, released tempests of ice and snowstorms with a simple flick of its goliath appendages. Aelar and his mates battled bravely, their powers uniting in an orchestra of ice and wrath against the gatekeeper's tireless surge.

Amidst the fight, Aelar encountered a significant association with Glacius. The watchman, when an exemplification of Eldrath's fury, uncovered the untold history of the Cryomancers and their harmonious relationship with the land. Aelar discovered that Eldrath itself was a conscious element, its frozen heart throbbing with the recurring pattern of sorcery.

With recently discovered understanding, Aelar and the Cryomancers fashioned a partnership with Glacius. The watchman, perceiving the reverberation between Aelar's embodiment and the Frostheart Precious stone, surrendered its guardianship. The gathering obtained the precious stone, its brilliant shine strengthening as it perceived the anointed one in Aelar.

The Frostheart Gem, presently in Aelar's control, gave to him a flood of force that rose above the limits of mortal comprehension. Aelar turned into a course of Eldrath's wizardry, his very presence entwined with the destiny of the frozen realm. The Cryomancers saw the satisfaction of the prescience — the Frozen Introduction had arrived at its peak.

As Aelar's power developed, so did the approaching danger that lingered over Eldrath. A break in the texture of reality showed, releasing an early stage force known as the Frostbane — an element brought into the world from the unrestrained turmoil of the frozen domains. The Frostbane looked to immerse Eldrath in a timeless winter, eradicating the actual pith of life that flourished inside.

Aelar, presently the harbinger of Eldrath's salvation, confronted the Frostbane in a fight that rose above the physical and the magical. The conflict of titanic powers resonated through the Frigid Fortress, deeply affecting its establishments. The Cryomancers, joined in reason, released their powers in an ensemble of ice and fire against the infringing haziness.

The fight seethed on, the actual texture of reality twisting and distorting as Aelar and the Frostbane took part in a dance of basic matchless quality. The skies above Eldrath turned into a kaleidoscope of shining lights and whirling storms, a demonstration of the inestimable battle that unfurled between the divinely selected individual and the early stage force that looked to consume everything.

Amidst the turmoil, Selene Frostheart, the Cryomancer Incomparable, joined the fight. Her dominance over ice was unmatched, and with a magnificent elegance, she added her powers to the invasion against the Frostbane. Aelar, perceiving Selene as both tutor and guide, felt a flood of solidarity that rose above the limits of his own capacities.

The perfection of the fight gravitated toward as Aelar, Selene, and the Cryomancers diverted their joined energies into a last, frantic attack. The Frostbane, squirming and reshaping, battled against the invasion of Eldrath's protectors. The very air popped with the force of their conflict, and the destiny of Eldrath remained in a precarious situation.

In a mental breakthrough, Aelar took advantage of the substance of Eldrath itself. He turned into a conductor for ice and snow, however for the very life force that beat through the frozen realm. With an otherworldly eruption of force, Aelar released a flood of warmth and essentialness that neutralized the Frostbane's cold grasp.

The early stage force withdrew, its malignant presence reducing as Aelar's brilliant energy wrapped Eldrath. The Cryomancers, washed in the luminosity of the stupendous fight, saw the resurrection of their reality. The timeless winter that had held Eldrath for quite a long time started to defrost, giving way to the commitment of another season.

As the skies cleared and the last remnants of the Frostbane scattered, Aelar remained at the cliff of Eldrath's change. The Frozen Preface had not exclusively been a harbinger of destruction yet an impetus for restoration. The Cryomancers, when limited by the frigid shackles of destiny, presently embraced the glow of a future liberated by the ghost of everlasting winter.

Eldrath, washed in the delicate shine of the arising sun, demonstrated the veracity of another time. The Cold Bastion remained as a demonstration of the flexibility of its kin, the translucent towers shimmering with the expectation that had been encouraged in the cauldron of difficulty. Aelar, having satisfied his predetermination, turned into an image of solidarity — an extension between the natural powers that molded Eldrath's fate.

Selene Frostheart, her eyes mirroring the pride and insight that accompanied age, moved toward Aelar. The Cryomancer Incomparable recognized the youthful legend, perceiving the extraordinary power he had saddled. Aelar, lowered by the greatness of his excursion, watched out at the tremendous spread of Eldrath, realizing that the Frozen Preface had changed the actual texture of their reality.

In the repercussions of the climactic fight, the Cryomancers devoted themselves to reconstructing Eldrath. The defrosted scenes, when frozen in time, overflowed with life as vegetation thrived in the recently discovered warmth. Towns that had endured the everlasting winter presently clamored with action, and chuckling reverberated through the once-quiet hallways of the Frosty Stronghold.

The solidarity fashioned in the pot of misfortune persevered as the Cryomancers, directed by Aelar and Selene, worked eagerly to fortify the connections between their natural powers and the land they called home. Eldrath, as of now not limited by the steady grasp of winter, prospered under the consideration of its versatile occupants.

As the seasons spun through their immortal dance, Eldrath embraced a congruity that rose above the ages. The Frozen Preface, when a prescience of destruction, had turned into a story of resurrection and recharging. The Cryomancers, watchmen of Eldrath's heritage, stood cautious, their association with the components solid.

In the core of Eldrath, Aelar, presently a respected figure, pondered the excursion that had formed his predetermination. The Frozen Introduction had been a preface not exclusively to calamity yet in addition to the strength of the human soul and the persevering through force of solidarity. The reverberations of Eldrath's frozen past murmured in the breeze, an update that predetermination, similar to the steadily evolving seasons, was an embroidery woven by the hands of time.

Thus, in the land where winter's hug was everlasting, Eldrath embraced a future washed in the glow of another sunrise — a demonstration of the dauntless soul that had endured the hardships of destiny. The Frozen Preface had played out its orchestra, and Eldrath remained as a living demonstration of the getting through force of trust despite an unsure fate.

1.1 Introduction to the world where temperature holds a mysterious power.

In a world not at all like our own, where the actual texture of the truth is woven with strings of otherworldly energy, temperature holds a secretive power that resists the customary laws of nature. This domain, known as Thermodar, exists in a sensitive harmony between burning intensity and frigid cold, where the recurring pattern of temperature direct the course of life itself.

The place that is known for Thermodar is an embroidery of limits, a tremendous span where locales of oppressive intensity flawlessly change into frozen scenes with no advance notice. The skies above are painted in tints of searing reds and cold blues, a steady sign of the basic powers that shape the world. In Thermodar, the very air snaps with the idle energy of temperature, an energy saddled and employed by the occupants of this baffling domain.

At the core of Thermodar lies the Nexus, a point of convergence of natural power that oversees the fragile harmony of temperature across the land. The Nexus, an enormous construction of throbbing energy, is a nexus of actual topography as well as of the ethereal powers that tight spot the world together. It is said that the temperature of the Nexus decides the territory of Thermodar out of the blue, and the individuals who look to control its power hold the way in to the predetermination of the domain.

The occupants of Thermodar, known as Thermancers, are people gifted with the capacity to control temperature. These creatures, brought into the world with an association with the natural energies that penetrate their reality, can shape the very temperature around them, twisting it to their will. Thermancers are a different gathering, each receptive to a particular part of temperature control, going from the formation of burning heatwaves to the conjuring of frigid snowstorms.

In Thermodar, the social orders that have arisen are pretty much as different as the scenes they occupy. In the singing deserts of Ignis Sands, Thermancers outfit the force of burning intensity to shape giant glass structures that sparkle like jewels under the tireless sun. The occupants of Ignis Sands have excelled at fire control, utilizing it for viable purposes as well as a type of articulation, making perplexing figures that appear to hit the dance floor with the undulating influxes of intensity.

On the far edge of the range lies the cold field of Cryo Pinnacles, where Thermancers spend significant time in the control of frigid temperatures. Here, strong fortifications made of captivated ice stand as landmarks to the strength of the people who call this cold domain home. The Thermancers of Cryo Pinnacles have fostered a remarkable type of cryomancy, forming glacial masses and bringing snowstorms to guard their region against any who might look to infringe upon their frozen space.

Between these limits, immense calm zones like the Breeze Fields exist, where Thermancers employ an agreeable harmony between intensity and cold. In these districts, farming twists, and networks flourish under the cautious direction of Thermancers who can control temperature to guarantee abundant harvests and gentle environments. The Breeze Fields act as a blend of Thermancer societies, where experts from various temperature disciplines meet up in a sensitive dance of participation.

In spite of the variety of Thermancer social orders, there is a propensity of pressure that courses through the veins of Thermodar. The Nexus, with its consistently moving temperature, is a wellspring of both power and eccentricism. The people who try to control the Nexus compete for strength, their desires conflicting in a ceaseless battle for matchless quality. The temperature of the Nexus is a sought after prize, and the Thermancers, each with their exceptional capacities, assume a crucial part in the unfurling show of Thermodar.

Amidst this unstable scene, a figure known as Aurelia Emberfrost arises — a Thermancer of exceptional expertise and desire. Aurelia, with hair on fire like gleaming blazes and eyes that shine with a frigid purpose, leaves on a mission to open the insider facts of the Nexus and outfit its temperature-controlling power for herself. Her process takes her across the singing deserts, through the frozen pinnacles, and into the calm heartlands, where she experiences difficulties that test the restrictions of her capacities.

Aurelia's odyssey carries her into contact with Thermancers from varying backgrounds. She frames unions with the people who share her vision and appearances enemies who might remain determined to control the temperature-employing force of the Nexus. The actual Nexus turns into a person in Aurelia's story, its temperature throbbing because of the unfurling occasions, a living epitome of the baffling powers that oversee Thermodar.

As Aurelia dives further into the core of Thermodar's secrets, she reveals old predictions that anticipate of a picked Thermancer who will join the different groups and carry equilibrium to the consistently moving temperatures of the domain. The predictions discuss an intermingling — a second when the Nexus' temperature will arrive at a crucial point, and the destiny of Thermodar will be chosen.

The excursion turns into a pot for Aurelia, a trial of her personality and capacities. En route, she learns not exclusively to dominate the control of temperature yet additionally to explore the complicated snare of unions and competitions that characterize Thermancer society. Aurelia finds that the ability to control temperature isn't simply a question of basic dominance yet an impression of the Thermancer's internal harmony, a sensitive harmony among enthusiasm and levelheadedness.

As Aurelia's journey moves toward its peak, the temperature of the Nexus turns out to be progressively temperamental. Heatwaves burn through Cryo Pinnacles, and snowstorms clear across Ignis Sands. The actual texture of Thermodar unwinds as the Nexus wavers near the precarious edge of turmoil. Aurelia understands that she remains at the junction of fate, her decisions reverberating through the natural flows that weave the embroidered artwork of Thermodar's presence.

In a climactic showdown at the Nexus, Aurelia faces foes who try to take advantage of the turmoil for their own benefit. The fight unfurls in an orchestra of basic rage, with temperatures wavering between limits. The very air pops with the power of Aurelia's battle, and the Nexus answers her internal agreement, its temperature settling under the Thermancer's undaunted control.

In the consequence of the contention, Aurelia remains before the Nexus as a signal of solidarity. The different groups of Thermancers, once in conflict, perceive the insight in her activities. The temperature of the Nexus sinks into an agreeable equilibrium, mirroring the newly discovered balance that Aurelia has brought to Thermodar. The predictions have been satisfied, and the picked Thermancer has arisen successful.

Thermodar, presently joined under Aurelia's initiative, enters another time of collaboration and understanding. The singing deserts and frozen tops become images of a common legacy, and Thermancers from all disciplines meet up to bridle the temperature-controlling power to improve their domain. The Nexus, when a wellspring of conflict, turns into an image of solidarity, its temperature mirroring the aggregate concordance of Thermodar's occupants.

As Aurelia expects a job of initiative, she directs Thermodar towards a future where temperature is as of now not a wellspring of division however a power that joins together. The Thermancer social orders, when disconnected in their basic territories, team up to open the maximum capacity of their capacities. New advances arise, mixing the singing intensity of Ignis Sands with the cold brightness of Cryo Tops in a blend of basic development.

The scenes of Thermodar change as the Thermancers cooperate to shape their reality. Urban communities of glass and charmed ice ascend starting from the earliest stage, as demonstration of the solidarity that Aurelia has cultivated. The Breeze Fields, when an impartial ground, become a flourishing center point of social trade, where Thermancers from all disciplines assemble to share their insight and encounters.

Aurelia's story turns into a legend went down through ages, an account of a Thermancer who transcended the difficulties of her opportunity to carry solidarity to Thermodar. The Nexus, presently a worshipped image, is monitored by Thermancers who guarantee that its temperature stays in ceaseless harmony.

The actual pith of temperature control turns into a sacrosanct workmanship, rehearsed not so much for strength but rather to improve Thermodar and its occupants.

Thus, in the existence where temperature holds a puzzling power, Aurelia Emberfrost's heritage perseveres. Thermodar, when a domain near the very edge of turmoil, flourishes in the amicable equilibrium that she accomplished. The recurring pattern of temperature, when a wellspring of disunity, turns into a demonstration of the flexibility of Thermodar's kin and the getting through force of solidarity notwithstanding basic limits.

1.2 Establish the setting, a society where people's emotions and abilities are influenced by temperature changes.

In the tremendous scope of the world Iridia, a domain formed by the whimsical dance of temperatures, a one of a kind society prospered where the recurring pattern of feelings and capacities were characteristically attached to the consistently moving environment. Here, the climate wasn't only a setting yet a living power that directed the actual embodiment of life. As the temperature vacillated, so too did the hearts and capacities of individuals who called this exceptional land home.

The place that is known for Iridia was a mosaic of different environments, from searing deserts heated by constant daylight to frozen tundras shrouded in never-ending ice. The skies above were a material of moving tones, reflecting the irregular disposition of the components. In Iridia, temperature wasn't simply a meteorological peculiarity; it was an impetus for the sign of phenomenal capacities and the regulation of significant feelings.

At the core of Iridia stood the Thermospire, a fantastic construction that filled in as the focal point of the domain's supernatural energies. This transcending building, throbbing with an ethereal shine, was accepted to be a channel between the actual world and the supernatural powers that represented the domain's environment. The Thermospire's temperature resolved the territory of Iridia, impacting the feelings and abilities of its occupants in manners both unobtrusive and significant.

Individuals of Iridia, known as Thermans, were sensitive to the temperature's impact from birth. Each Therman had a special capacity, an indication of their internal identities that came and went with the temperature. In the brutal intensity, some Thermans acquired the ability to control flares, while others wound up strengthened with an unrivaled readiness. Alternately, in the gnawing cold, capacities moved to cryomancy or the expansion of actual strength.

Feelings, as well, were dependent upon the temperature's impulses. In the glow of a calm day, euphoria and kinship bloomed among Thermans, encouraging a feeling of local area that rose above individual contrasts. On the other hand, when the temperature plunged, a peaceful despairing settled over the land, and thoughtfulness grabbed hold as Thermans wrestled with the grave magnificence of a snowy scene.

One Therman, named Elysia Emberfrost, ended up at the core of this many-sided dance among temperature and capacity. With hair as blazing as the midsummer sun and eyes that mirrored the cold profundities of winter, Elysia exemplified the double idea of Iridia's environment. Her capacity, similar to her demeanor, moved couple with the Thermospire's temperature, making her a living indicator of the domain's enchanted energies.

Elysia's process started in the city of Thermara, settled in a locale where the temperature cycled between limits. Thermara was a clamoring city with engineering that reflected the transient idea of its environment. Structures made from charmed materials that answered temperature changes decorated the scene, their varieties moving from warm shades to cool tones as the day advanced.

In Thermara, Elysia explored the clamoring roads, where Thermans of any age approached their day to day routines. Traders hawked products that reverberated with the common temperature, from cooling mixtures during boiling evenings to warming elixirs for crisp nights. The actual city appeared to be alive, answering the Thermospire's temperature vacillations with an orchestra of moving varieties and encompassing enchantment.

Elysia's capacity, a combination of pyromancy and cryomancy, made her a special figure in Thermara. As the Thermospire's temperature increased, blazes moved readily available, and warmth exuded from her actual presence. On the other hand, in the virus embrace of a colder time of year night, her touch could freeze the air, leaving fragile ice designs afterward. Elysia's duality made her both respected and dreaded, a living epitome of Iridia's perplexing nature.

As Elysia leveled up her skills, she found a baffling association between the Thermospire's temperature and her own feelings. The changes in her state of mind reflected the domain's environment, as though the actual pith of Iridia coursed through her veins. This acknowledgment drove her to set out on a journey to disentangle the insider facts of the Thermospire and grasp the significant exchange between temperature, feelings, and capacities.

Her process took her across Iridia's assorted scenes, from the sizzling deserts of Coal Rises to the glasslike pinnacles of Frostfall Reach. En route, she experienced Thermans who, similar to her, wrestled with the mind boggling equilibrium of their own capacities and feelings. Some had figured out how to bridle the temperature's impact with accuracy, while others battled against the erratic idea of their powers.

In Ash Hills, where searing heatwaves could diminish the hardest stone to liquid slag, Elysia experienced a Therman named Rurik Sandsoul. Rurik, with skin tanned by years under the persistent sun, had excelled at controlling sand with the accuracy of a stone worker. His capacities permitted him to summon twirling dust storms that both secured and darkened, mirroring the brutal excellence of the desert.

As Elysia and Rurik shared their encounters, they found a consistent theme that associated their excursions. Both wrestled with the duality of their capacities, looking for a congruity that would permit them to employ the temperature's impact without being consumed by its limits. Together, they wandered further into the secrets of Iridia, their bond developing further as they defied the difficulties that lay ahead.

Their mission drove them to Frostfall Reach, where the air was bone chilling to the point that even breath could solidify. In the midst of the frozen scenes, Elysia and Rurik experienced a Therman named Lyra Frosthaven. Lyra, with hair as white as newly fallen snow, been able to gather mind boggling ice molds that sparkled like precious stones in the evening glow. Her powers mirrored the peaceful magnificence of a colder time of year's evening.

Lyra, similar to Elysia and Rurik, wrestled with the close to home cost of her capacities. In the isolation of Frostfall Reach, where the quietness was broken exclusively by the mash of snow underneath their boots, the triplet shared accounts of their battles and wins. Through these common encounters, they fashioned a bond that rose above the limits of their singular capacities, framing a trinity that repeated the sensitive equilibrium of Iridia itself.

As Elysia, Rurik, and Lyra dove further into their journey, they uncovered old texts that discussed a forecasted occasion — the Intermingling of Components. As per these texts, when the Thermospire's temperature arrived at a basic point, Thermans who typified the domain's different environments would meet up to open a power that could reshape Iridia's fate.

The Thermans understood that the Assembly of Components was drawing closer, and they felt the Thermospire's temperature fluctuating with expanding power. Elysia, Rurik, and Lyra, presently joined by Thermans from various districts, remained at the incline of a disclosure. The Thermospire's temperature, impacted by the aggregate feelings of its occupants, held the way to opening a likely that rose above individual capacities.

In the climactic snapshot of the Combination of Components, the Thermans assembled at the Thermospire, their joined presence reverberating with the enchanted energies that saturated Iridia. The temperature of the Thermospire flooded, answering the feelings flowing through the Thermans' hearts. Blazes and ice entwined in an entrancing presentation, projecting a brilliant gleam across the scene.

At that time of solidarity, Elysia, Rurik, and Lyra felt a flood of force not at all like anything they had encountered previously. The limits between their capacities obscured as they tackled an agreeable combination of fire, ice, and sand. The very air around them vibrated with the reverberation of the Union, and the Thermospire's temperature settled in a condition of wonderful balance.

The Thermans, their capacities currently enhanced by the Assembly, understood that Iridia itself had answered their aggregate agreement. The domain, when portrayed by the capricious dance of temperature, presently embraced a newly discovered balance. The searing deserts, frozen tops, and mild scenes coincided in a fragile congruity that reflected the solidarity manufactured by the Thermans.

Elysia, Rurik, and Lyra, presently injected with the force of the Assembly, turned into the watchmen of Iridia's balance. They cooperated to guarantee that the Thermospire's temperature stayed in wonderful equilibrium, impacting the domain's environment in manners that cultivated development, success, and solidarity. The once-disparate Thermans presently teamed up to share their insight and encounters, rising above the restrictions forced by their singular capacities.

Iridia, when a domain of limits, flourished in the fallout of the Union of Components. Urban areas thrived, decorated with charms that commended the variety of temperature impacts. Thermans, when characterized by the locales they hailed from, presently embraced an aggregate personality that praised the agreeable exchange of fire, ice, and sand.

Elysia, Rurik, and Lyra, presently venerated as the Natural Set of three, directed Iridia into another time of success and collaboration. The Thermospire remained as a demonstration of the Thermans' capacity to rise above the intrinsic difficulties of their one of a kind society. The temperature, when a power that directed the back and forth movement of feelings and capacities, turned into a wellspring of motivation — an indication of the extraordinary power that lay inside solidarity and equilibrium.

Thus, in the domain of Iridia, where temperature held a secretive power, the Combination of Components turned into an image of trust and recharging. The Thermans, joined by their common encounters and the concordance of the Intermingling, left on an excursion of investigation and disclosure. The once mysterious dance of temperature turned into a festival of Iridia's variety.

1.3 Introduce the protagonist, a young individual discovering their unique connection to temperature.

In a curious town settled between moving slopes and wandering streams, a youthful individual named Alaric Grayson wound up on the cusp of a remarkable disclosure. Alaric, with disheveled chestnut hair and eyes that reflected the profundities of the unmistakable summer sky, carried on with an apparently customary life in the town of Eldenbrook. However, as the sun plunged underneath the skyline one critical night, Alaric's reality would be perpetually different by a disclosure that rose above the limits of the standard.

Eldenbrook, washed in the warm shades of dusk, radiated an ideal appeal that gave a false representation of the secrets hid inside its environmental factors. Ages of families had called this town home, their lives woven into the texture of an affectionate local area. Eldenbrook was known for its dynamic celebrations, exuberant business sectors, and the deep rooted custom of passing down stories of old predictions that murmured of stowed away powers.

As the night breeze helped the fragrance of sprouting wildflowers through the air, Alaric meandered along the town edges, enthralled by the excellence of the blurring light. The town older folks frequently discussed a special association between Eldenbrook's occupants and the temperature that represented the land. Alaric, in any case, had consistently viewed these stories as fables, a frivolity to add tone to the embroidery of town life.

The snapshot of disclosure happened in a separated dale washed in the delicate gleam of fireflies. Alaric, drawn by an odd power, coincidentally found a secret clearing where an old stone platform stood, decorated with otherworldly images scratched into its endured surface. As Alaric drew nearer, an ethereal reverberation reverberated through the dell, an agreeable murmur that appeared to answer the actual heartbeat of the land.

Captivated, Alaric set a hand upon the stone platform, and at that time, a flood of warmth coursed through their veins. It was as though the actual embodiment of Eldenbrook's temperature had recognized Alaric's presence. The images on the platform shined with a powerful light, projecting perplexing examples that reflected the star groupings above.

As the acknowledgment unfolded upon Alaric, the association with temperature uncovered itself in a fountain of tangible encounters. The glow that coursed through Alaric's fingertips turned into their very own expansion being, a freshly discovered mindfulness that rose above the common limits of touch. The air around Alaric gleamed with a concealed energy, and the encompassing temperature appeared to answer the implicit association that presently bound them to the land.

In the days that followed, Alaric investigated this freshly discovered association with a feeling of miracle and fear. Eldenbrook, when natural, presently felt like a domain of vast conceivable outcomes. Alaric could detect the temperature's vacillations with an intrinsic accuracy, receptive to the inconspicuous movements that represented the town and its environmental factors. The glow of the sun, the cool wind of the night, and the delicate dash of downpour — all became tangible strings in the complex woven artwork of Alaric's presence.

The town older folks, perceiving the meaning of Alaric's disclosure, directed the youthful person through the antiquated lessons went down through ages. Alaric discovered that the locals of Eldenbrook were not simple spectators of the temperature's impulses but rather dynamic members in a cooperative relationship with the land. Every resident, it appeared, had a one of a kind association with temperature, an otherworldly bond that appeared in different ways.

Alaric's capacities started to appear as an unobtrusive effect on the nearby environment. On warm evenings, the sun appeared to sparkle a piece more splendid when Alaric was close, and blossoms sprouted in energetic tints as though answering a quiet order. On the other hand, during cool nights, a delicate breeze would go with Alaric's presence, conveying with it a feeling of quiet and serenity.

The town, once uninformed about the phenomenal idea of its occupants, presently wondered about Alaric's capacities. Murmurs of a favored one, predicted in the old predictions, spread through Eldenbrook like quickly. Alaric, nonetheless, stayed humble despite newly discovered consideration, wrestling with the obligations that accompanied their special association with temperature.

In the core of Eldenbrook, a social occasion was gathered in the town square. The elderly folks, enhanced in formal robes that bore images repeating those on the old platform, remained before the collected townspeople. Alaric, directed by a combination of interest and fear, moved toward the social occasion as the mumbles of the group slowly quieted into an eager quietness.

The town senior, a shrewd figure named Elara Whisperwind, discussed the old predictions that had predicted the appearance of a Thermancer — a Therman with a special association with temperature who might assume a significant part in Eldenbrook's fate. The older folks, perceiving Alaric as the anointed one, uncovered the meaning of the Thermancer's job in keeping up with the sensitive equilibrium of temperature that supported the town and its occupants.

Alaric, presently mindful of the heaviness of their newly discovered predetermination, left on an excursion of disclosure directed by the town elderly folks. The Thermancer's way elaborate excelling at temperature control, an errand that required adjusting not exclusively to the outside environment yet additionally to the inner rhythms of feelings and purpose.

The elderly folks acquainted Alaric with the Basic Forest, a holy spot concealed profound inside Eldenbrook's heart. This supernatural woods, washed in the dappled light of old trees, filled in as a preparation ground for Thermancers of previous eras. Here, Alaric experienced the substance of temperature in its most perfect structure — an ensemble of warmth and coolness, a dance of fire and ice that resounded with the actual soul of Eldenbrook.

Under the elderly folks' direction, Alaric rehearsed temperature control, leveling up the skill to impact the environment as one with the town's requirements. From the beginning, the indications were inconspicuous — a delicate breeze to cool a warm summer day or an eruption of warmth to fight off a startling chill. As Alaric's abilities advanced, so did the size of their impact over Eldenbrook's temperature.

The town, when dependent on the impulses of nature, presently encountered a recently discovered harmony under Alaric's cautious stewardship. Crops prospered with an accuracy that reflected the Thermancer's plan, and Eldenbrook's environment turned into a demonstration of the congruity between occupants and the temperature embraced them.

However, the excursion of a Thermancer was not without challenges. Alaric experienced snapshots of inward conflict, where feelings took steps to disturb the sensitive equilibrium of temperature control. In the midst of pain, Eldenbrook's environment reflected Alaric's unrest, sending waves of lopsidedness through the town.

It was during one such testing second that Alaric found a surprising partner — an individual resident named Seraphina. Seraphina, with emerald-green eyes that shimmered sincerely, had consistently respected Alaric's capacities from a good ways. Detecting Alaric's internal battle, Seraphina drew closer with sympathy and understanding, offering a point of view that resounded with the Thermancer's own excursion.

Together, Alaric and Seraphina explored the intricacies of temperature control. Seraphina, however not a Thermancer herself, had an intrinsic insight that supplemented Alaric's capacities. The two produced an association that reached out past the domain of temperature, a bond established in shared encounters and a common longing to protect Eldenbrook's thriving.

As Alaric and Seraphina dug into the secrets of temperature control, they revealed an old relic — the Emberfrost Emblem. The emblem, enhanced with images suggestive of those on the stone platform, was said to intensify a Thermancer's capacities and act as a point of convergence for diverting temperature-related energies.

The revelation of the Emberfrost Emblem denoted a defining moment in Alaric's excursion. With the emblem's direction, Alaric and Seraphina wandered into the Natural Forest, looking to open the maximum capacity of the Thermancer's capacities. As they remained inside the forest's consecrated limits, a reverberation reverberated through the air, and the emblem beat with a brilliant energy that reflected the heartbeat of Eldenbrook.

Alaric, directed by the emblem's power, encountered a flood of warmth that rose above the actual domain. Blazes moved readily available, and the actual air appeared to answer Alaric's purpose, adapting to the Thermancer's will. The emblem's power permitted Alaric to wind around complicated examples of temperature control, impacting the environment with an accuracy that outperformed anything encountered previously.

The town elderly folks, seeing Alaric's freshly discovered dominance, recognized the Thermancer's development with a grave service. Alaric was decorated with the Emberfrost Emblem, an image of the association between Eldenbrook's past and its promising future.

The town, when dependent on the flighty dance of nature, presently remained under the vigilant look of a Thermancer who had bridled the temperature's power.

With the Emberfrost Emblem as a point of convergence, Alaric and Seraphina cooperated to grow the limits of Eldenbrook's flourishing. The town turned into a sanctuary where temperature and feeling moved couple, making a climate of congruity that resounded with the actual embodiment of their special association with the land.

As seasons spun through their ageless dance, Eldenbrook flourished under Alaric's stewardship. The harvests yielded plentiful harvests, and the town turned into a reference point of flourishing in the core of Iridia. Alaric's excursion, when a disclosure of individual capacities, had changed into an adventure that formed the fate of Eldenbrook and its occupants.

In the nightfall of a fall evening, as Eldenbrook arranged for its yearly celebration, Alaric remained at the focal point of the town square. The Emberfrost Emblem sparkled with an internal fire, reflecting the glinting lights that decorated the environmental elements. Locals, youthful and old, accumulated to observe the perfection of Alaric's excursion — an excursion that had uncovered the significant association between temperature, feelings, and the strength of the human soul.

The celebration unfurled in a scene of light and sound, a festival of Eldenbrook's one of a kind relationship with temperature. Alaric, presently a venerated figure, imparted the stage to Seraphina, whose insight and compassion played had an instrumental impact in the Thermancer's development. Together, they exhibited the amicable dance of temperature control, a presentation that spellbound the hearts of Eldenbrook's occupants.

As the celebration arrived at its peak, Alaric stretched out appreciation to the town seniors, the companions who had remained close by, and the place that is known for Eldenbrook itself. The Emberfrost Emblem, its brilliant sparkle projecting a warm emanation over the town square, represented the solidarity among temperature and the unyielding soul of its kin.

Thus, in the town of Eldenbrook, where temperature held a secretive power, Alaric Grayson's process turned into a demonstration of the groundbreaking capability of disclosure, flexibility, and the significant association between an individual and the land they called home. The glow that moved through Alaric's veins repeated the heartbeat of Eldenbrook, a heartbeat that beat with the commitment of a future molded by the fragile dance of temperature and the getting through soul of its occupants.

Chapter 2

The Thermodynamic Gift

In the city of Thermasburg, where transcending towers of captivated gem scratched the sky and obscure energy beat through the air, carried on with a youthful individual named Elinor Veridian. Elinor, with her lively emerald eyes and a fountain of reddish-brown twists, had an exceptional gift that put her aside in a general public where wizardry and innovation met in a stunning showcase of development. Her gift was known as the Thermodynamic Gift, a capacity to control temperature and saddle the inactive energy that pervaded the city.

Thermasburg, a wonder of both magical and innovative ability, flourished with the fragile harmony between the basic powers that represented its presence. The city's establishments were established in the union of intensity and cool, an orchestra of temperatures that controlled the complex hardware and supported the charmed designs. The residents of Thermasburg, known as Thermans, were adroit at diverting these temperature changes for different purposes, from driving their homes to powering the hypnotizing presentations of esoteric lights that embellished the cityscape.

Elinor, notwithstanding, had a Thermodynamic Gift that outperformed the conventional capacities of her kindred Thermans. Her touch could incite a controlled flood of nuclear power, making pockets of extreme intensity or chilling ice readily available. This exceptional ability had appeared in Elinor since early on, a strange legacy from her precursors that set before her a way of both interest and disclosure.

As Elinor developed, so did how she might interpret the Thermodynamic Gift. She figured out how to employ the temperature variances as a wellspring of force as well as a type of self-articulation. In Thermasburg's eminent foundations, where Thermans improved their abilities in the hidden expressions and mechanical authority, Elinor's capacities collected both esteem and interest.

Her tutor, Teacher Octavius Flameforge, a recognized Therman with a mane of red hot hair and an insightful disposition, perceived the expected inside Elinor's Thermodynamic Gift. Under his direction, Elinor dove into the mind boggling hypotheses that supported temperature control, investigating the sensitive harmony among intensity and cold, entropy and request.

Thermasburg, regardless of its mechanical wonders, confronted difficulties that undermined the balance of its charmed biological system. Erratic changes in temperature had prompted disturbances in the city's power framework, and peculiarities in the hidden energies had brought about peculiarities that opposed clarification. Elinor, driven by a feeling of obligation and a craving to open the maximum capacity of her gift, willingly volunteered to investigate the secrets that covered Thermasburg's fragile equilibrium.

As Elinor dug into her examination, she revealed old texts that discussed an incredible relic — the Embercore Crystal. As indicated by the texts, the Embercore Crystal had the capacity to intensify a Therman's command over temperature, offering a degree of dominance that rose above common limits. The antiquity was supposed to be concealed in the core of Thermasburg, monitored by enchanted wards and the reverberations of old gatekeepers.

Not entirely settled to bridle the force of the Embercore Crystal, Elinor left on a journey that drove her through the overly complex halls of Thermasburg's obscure chronicles and into the neglected chambers where the city's set of experiences lay laced with the actual texture of its presence. En route, Elinor experienced difficulties that tried the restrictions of her Thermodynamic Gift, from burning chambers where the intensity took steps to overpower her to frozen chambers where the unpleasant virus looked to douse her inward fire.

As Elinor explored the secret openings of Thermasburg, she revealed an organization of old courses that diverted the city's temperature energies. The courses, throbbing with a musical murmur, appeared to answer Elinor's Thermodynamic Gift. With each controlled flood of temperature, she enacted torpid systems and disentangled the layers of security encompassing the Embercore Crystal.

The excursion, in any case, was not without its hazards. Elinor confronted preliminaries that tried her capacities as well as how she might interpret the sensitive equilibrium inside Thermasburg. She experienced rival Thermans who looked for the force of the Embercore Crystal for themselves, each determined by desires that undermined the actual strength of the city.

Among these opponents was a charming Therman named Viktor Drakon, whose silver hair and puncturing blue eyes concealed a tireless assurance to bridle the Embercore Crystal's power for his vision of Thermasburg's future. Viktor, with his dominance of both fire and ice control, turned into a considerable foe in Elinor's journey.

The conflict among Elinor and Viktor reverberated through the old offices of Thermasburg, a fight that interweaved the dance of temperature with the ensemble of esoteric energies. Flares moved as one with ice as the Thermans released their

Thermodynamic Gifts, each trying to acquire the high ground and guarantee the Embercore Crystal for their own.

In the climactic showdown, Elinor's natural comprehension of the sensitive equilibrium inside Thermasburg ended up being her most noteworthy strength. She saddled the temperature vacillations with an artfulness that reflected the city's interconnected frameworks, countering Viktor's forceful methodology with a deliberate control that addressed the quintessence of her Thermodynamic Gift.

As the Embercore Crystal's defensive wards faltered under the invasion of contending temperatures, Elinor immediately jumping all over the chance to adjust herself to the curio's lethargic energies. The Crystal, reverberating with the exceptional mark of her Thermodynamic Gift, answered her touch, and a flood of force flowed through Elinor's veins.

At that time of association, Elinor witnessed the huge possible that lay inside the Embercore Crystal. The relic, presently in her control, transmitted with an internal splendor, its multi-layered surfaces catching the embodiment of Thermasburg's temperature energies. The city, when near the very edge of unsteadiness, appeared to answer the Embercore Crystal's enactment, subsiding into an agreeable balance.

Viktor, lowered by the acknowledgment that the Embercore Crystal had picked Elinor as its wielder, hesitantly recognized her authority. The opponents, when secured in a fight for predominance, presently stood together in the core of Thermasburg, a city that demonstrated the veracity of the union of fire and ice, innovation and sorcery.

With the Embercore Crystal as her aide, Elinor turned into a signal of soundness inside Thermasburg. The antiquity permitted her to channel temperature energies with extraordinary accuracy, guaranteeing the city's thriving and shielding it from the erratic oddities that had once undermined its sensitive equilibrium.

Elinor's excursion, from a talented Therman to the wielder of the Embercore Crystal, turned into a legend inside Thermasburg. Her story reverberated through the translucent passages and captivated towers, a demonstration of the extraordinary force of understanding and dominating the Thermodynamic Gift. The city, presently washed in the brilliant shine of Elinor's impact, entered another time of development and thriving.

As the years passed, Elinor expected a job of authority inside Thermasburg, directing the Thermans in their investigation of temperature control and the mix of wizardry and innovation. The Embercore Crystal, cherished as an image of Thermasburg's flexibility, filled in as a point of convergence for collective get-togethers and festivities, a sign of the sensitive equilibrium that characterized the city's presence.

In the core of Thermasburg, where temperature held a puzzling power, Elinor Veridian's heritage persevered. The city, when near the precarious edge of wild change, flourished under her vigilant look and the brilliant impact of the Embercore Crystal. The Thermodynamic Gift, when a wellspring of both marvel and challenge, turned into a wellspring of motivation for Thermans who tried to investigate the limits of their own capacities and the complex dance of temperature that molded their reality.

2.1 Explore the protagonist's discovery of their ability to control and manipulate temperature.

In the little beach front town of Eldermere, settled between rocky bluffs and the field of the ocean, carried on with a youthful individual named Aria Wintergale. Aria, with her striking sapphire eyes and an outpouring of 12 PM dark hair that outlined her face, had an apparently common existence locally that flourished with the recurring pattern of the tides. Much to her dismay that the everyday cadence of day to day presence was going to be disturbed by a disclosure that would show her a way of self-revelation and uncover a power lethargic inside her — the capacity to control and control temperature.

Eldermere, however interesting, held a specific sorcery inside its beach front hug. The town's appeal exuded from the exchange of the sea's flows, the delicate influence of ocean growth in the tide, and the delicate murmurs of the breeze that conveyed stories from far off shores. Local people, focused anglers and craftsmans, delighted in the effortlessness of their lives, uninformed about the idle likely that waited underneath the surface.

Aria, a curious soul with a voracious interest, frequently ended up attracted to the edge of the precipices, where the pungent breeze kissed her cheeks and the cadenced sound of waves running into the stones swirled all around. One pivotal night, as the sun plunged underneath the skyline, projecting the sky in tones of orange and pink, Aria encountered an exceptional sensation — a glow that radiated from the center of her being.

Interested by this newly discovered warmth, Aria expanded her hands towards the ocean breeze. To her shock, the air appeared to answer her aim. A delicate expansion in temperature went with her signal, making a pocket of warmth that diverged from the cool sea breeze. Bewilderment moved quickly over Aria's face as she explored different avenues regarding this abnormal peculiarity, understanding that the temperature around her answered her implicit cravings.

As days transformed into weeks, Aria's capacity to control temperature advanced. She found that her feelings assumed a critical part in this freshly discovered power. At the point when euphoria made her exuberantly pleased, an unobtrusive warmth encompassed her environmental elements, making a consoling feeling. On the other hand, snapshots of thoughtfulness and serenity delivered a cooling sensation, as though the actual air reflected the quietness inside Aria's spirit.

The acknowledgment of her interesting gift sent Aria on an excursion of self-disclosure. In the mystery of her everyday existence, she explored different avenues regarding the degree of her capacities. A straightforward touch could permeate an item with warmth, and an engaged idea could invoke a cool wind on a warm day. Aria, when a normal inhabitant of Eldermere, presently held a power that associated her to the very components that characterized her seaside home.

Expression of Aria's capacities started to spread through Eldermere, energized by the wonderment and interest of the individuals who saw her accomplishments. Murmurs of a skilled soul who could twist the temperature to her will circumnavigated through the very close local area, arriving at the ears of Eldermere's shrewd senior, Maevra Seastorm.

Maevra, a lady with silver hair that reflected the twilight waves and eyes that shone with old insight, perceived the meaning of Aria's recently discovered gift. She moved toward the youthful hero with a knowing grin, offering direction and bits of knowledge into the otherworldly idea of temperature control.

Under Maevra's tutelage, Aria discovered that Eldermere, regardless of its outward straightforwardness, held a remarkable energy — an amicable harmony between the ocean, the breeze, and the temperature that represented the town's environment. Maevra discussed an old heredity of gatekeepers in Eldermere, people who had the capacity to collective with the components and keep up with the fragile balance that supported the town.

Aria, it appeared, had a place with this genealogy. Her revelation of temperature control denoted the enlivening of a tribal power that lay torpid inside her bloodline. As she dove into the lessons passed somewhere around Maevra, Aria revealed the rich history of Eldermere and the significant association between gatekeepers and the natural powers formed their lives.

The town, it showed up, had confronted difficulties in the past that tried the strength of its watchmen. Irregularities in temperature, storms that opposed regular examples, and disturbances in the rhythmic movement of the tides — these were the preliminaries that the watchmen had explored through the ages. Aria, presently the most recent in this line of watchmen, conveyed the obligation of protecting Eldermere's fragile equilibrium.

As Aria improved her skills under Maevra's careful attention, she found the interconnectedness of temperature with the feelings and plan of the individuals who occupied Eldermere. The town's environment turned into a material whereupon the aggregate state of mind of its occupants was painted. A cheerful festival could appear as a delicate warmth, while a dismal second could summon a cool wind that reflected the thoughtfulness in the hearts of Eldermere's occupants.

However, the recently discovered familiarity with her obligations delivered difficulties for Aria. The sensitive equilibrium that she tried to keep up with was powerless to outer impacts, and Aria wrestled with snapshots of uncertainty and vulnerability. Her feelings, when a wellspring of force, turned into a potential disruptor of Eldermere's harmony.

In one such testing second, a furious tempest cleared over Eldermere, carrying with it violent breezes and heavy downpour. Aria, feeling the heaviness of her job as a gatekeeper, attempted to hold the disturbance inside her own feelings. The tempest, energized by the disunity in her heart, seethed with a force that Eldermere had not seen in ages.

Maevra, detecting Aria's unseen conflict, offered direction amidst the tempest. She talked about the significance of embracing one's feelings, recognizing them without being consumed by their storm. Aria, with assurance in her eyes, faced the profound whirlwind inside her, looking for an amicable goal that would conciliate the components she presently controlled.

In an epiphany, Aria directed her feelings into an engaged goal. The tempest, when a sign of inner unrest, answered her newly discovered balance. Winds died down, downpour changed into a delicate sprinkle, and the temperature around Eldermere balanced out. Aria, with a significant feeling of understanding, understood that her excursion as a watchman included the control of temperature as well as the dominance of her own profound storms.

Eldermere, presently quieted by Aria's freshly discovered understanding, embraced a peaceful serenity. The townsfolk, thankful for the arrival of harmony, respected Aria with a combination of deference and regard. The tempest had uncovered the extent of Aria's capacities as well as the versatility of her soul despite misfortune.

As seasons burned through their ageless dance, Aria kept on developing into her job as Eldermere's gatekeeper. How she might interpret temperature control extended past the prompt environment, enveloping the many-sided dance among feelings and the components.

The town, when a straightforward waterfront sanctuary, turned into a demonstration of the harmonious connection between watchman and the captivating powers characterized its presence.

In the core of Eldermere, where temperature held a puzzling power, Aria Wintergale's process unfurled as an embroidery woven with strings of disclosure, obligation, and self-acknowledgment. The town, presently receptive to the back and forth movement of Aria's impact, flourished in the delicate hug of its gatekeeper's freshly discovered concordance. The waves kept on running into the bluffs, the breeze murmured mysteries from far off shores, and the temperature answered the aggregate heartbeat of Eldermere — a town where the components moved in solidarity with the watchman who held their sensitive equilibrium in her skilled hands.

2.2 The protagonist meets a mentor or guide who helps them understand the significance of their power.

In the core of the old city of Luminastra, where transcending towers of translucent designs sparkled in the timeless shine of brilliant circles, carried on with a youthful individual named Orion Sunshard. Orion, with hair the shade of liquid gold and eyes that reflected the brightness of the heavenly circles above, had a remarkable gift — an association with the radiant energy that saturated Luminastra. Little did Orion had any idea that his experience with a secretive tutor would divulge the genuine meaning of his power and set him on an excursion of infinite extents.

Luminastra, a wonder of both heavenly wizardry and mechanical marvels, flourished with the iridescent energies outfit from the universe. The city's occupants, known as Luminars, were adroit at diverting this enormous energy for different purposes, from enlightening the cityscape to controlling high level apparatus that rose above the limits of traditional comprehension.

Orion, however mindful of the unprecedented idea of Luminastra, accepted his job as a Luminar was restricted to the commonplace errands relegated to him in the divine studios. One portentous evening, as Luminastra washed in the ethereal gleam of 1,000 stars, Orion encountered a flood of grandiose energy that flowed through his veins. It was as though the actual embodiment of the divine circles had recognized his presence, injecting him with a brilliant power that outperformed the normal capacities of his kindred Luminars.

Fascinated and fairly dumbfounded by this grandiose arousing, Orion looked for replies inside the divine files of Luminastra. The files, a vault of old insight and vast information, held the insider facts of the city's association with the divine domains. It was during his single investigation of these documents that Orion's way met with that of a puzzling figure — a sage embellished in streaming robes that reflected the heavenly bodies and eyes that shimmered with vast understanding.

The sage presented themselves as Seraphis Astralwind, a gatekeeper of Luminastra and guardian of heavenly information. Seraphis, with an atmosphere that resounded with the energies of the universe, detected the blossoming power inside Orion and perceived the meaning of his association with the glowing powers that administered the city.

Under Seraphis' direction, Orion left on an excursion of vast investigation. The wise discussed a heavenly peculiarity known as the Astral Combination — an arrangement of grandiose energies that happened once in a thousand years. During this intriguing combination, Luminars with a novel proclivity for heavenly powers could open the genuine capability of their powers and fashion an association with the actual texture of the universe.

Orion, presently mindful of the uncommon idea of his capacities, prepared under Seraphis' careful focus. The sage uncovered that Luminastra's glowing energy was not only an asset for commonsense applications but rather an impression of the city's significant association with the divine domains. Luminars, with their vast gifts, assumed a urgent part in keeping up with the fragile harmony between the heavenly powers and Luminastra's agreeable presence.

As Orion dove into his preparation, he found that his powers reached out past the control of brilliant energy. His touch could reverberate with the divine frequencies, opening secret pathways that rose above the actual domain. The very air around him sparkled with brilliant shades, answering the infinite reverberation that exuded from his being.

Amidst his preparation, Orion experienced snapshots of heavenly attunement. During these otherworldly episodes, he could see the inestimable embroidery that wove through the texture of Luminastra — an embroidered artwork that associated the city to far off cosmic systems, star groups, and domains outside human ability to grasp. It was a disclosure that extended's comprehension Orion might interpret his job as a Luminar and the grandiose powers that molded his fate.

Seraphis, with an insight sharpened by hundreds of years of inestimable fellowship, directed Orion through heavenly ceremonies and contemplations. They talked about the astral reverberations that resounded through Luminastra, conveying murmurs from the far off corners of the universe. Orion figured out how to explore these reverberations, knowing the vast messages that held the keys to opening the maximum capacity of his divine gift.

As Orion's preparation advanced, Luminastra confronted a divine inconsistency — a looming Astral Intermingling that blended the radiant energies to uncommon levels. The divine spheres, ordinarily a wellspring of direction and light, gleamed with a flighty splendor that reflected the vast unrest. Seraphis, detecting the desperation of the circumstance, rushed Orion's preparation, setting him up to embrace the difficulties that the approaching union would bring.

The Astral Union showed up in a blast of divine magnificence. Luminastra's skies turned into a kaleidoscope of shining lights, each circle transmitting with a force that rose above the limits of predictability. Luminars accumulated in wonderment and fear as the divine energies flooded through the city, making an ethereal feeling that resounded with the enormous powers at play.

Orion, directed by Seraphis, ventured into the core of the Astral Intermingling. The radiant energy answered his presence, encompassing him in a divine hug. In that extraordinary second, Orion's faculties extended past the limits of the actual domain. He saw the divine flows that moved through Luminastra, associating him to the actual embodiment of the universe.

The assembly, however sublime, additionally delivered heavenly difficulties. An astral crack showed, taking steps to upset the fragile harmony among Luminastra and the heavenly domains. Seraphis, presently uncovered as a Divine Gatekeeper endowed with keeping up with this equilibrium, encouraged Orion to channel his freshly discovered powers and seal the fracture before it released inestimable bedlam upon the city.

Orion, filled by the iridescent energies of the Astral Combination, embraced his job as Luminastra's gatekeeper. His touch resounded with the astral frequencies, and he expanded his hands towards the fracture. A flood of heavenly power exuded from him, winding through the brilliant energies and fixing the astral break. Luminastra, washed in the consequence of the union, stood tough against the divine difficulties that had compromised its presence.

The Luminars, seeing Orion's dominance over the vast powers, respected him with a combination of love and appreciation. The city, once defenseless against the impulses of divine inconsistencies, presently flourished under the direction of a manufactured a significant watchman association with the universe. Seraphis, playing satisfied their part as a tutor and guide, presented to Orion the title of Astral Superintendent — a divine steward depended with the continuous concordance among Luminastra and the heavenly domains.

Orion's excursion, from a common Luminar to the Astral Superintendent, turned into an adventure murmured through the radiant roads of Luminastra. The city, presently enlightened by its captivated circles as well as by the grandiose energies diverted by its gatekeeper, entered another time of heavenly success.

As seasons burned through their enormous dance, Luminastra embraced the divine inheritance produced by Orion Sunshard. The brilliant energies that flowed through the city reverberated with the astral reverberations, making a climate where the heavenly and the ordinary existed as a lovely, unified whole. Orion, with the insight offered to him by the Astral Union, directed Luminastra through the heavenly flows, guaranteeing the city's persevering through association with the universe.

In the core of Luminastra, where enormous energies held a secretive power, Orion Sunshard's process unfurled as a demonstration of the extraordinary capability of heavenly fellowship, grandiose comprehension, and the significant association between a gatekeeper and the glowing city they called home. The splendor of the heavenly spheres reflected the brilliance of Luminastra's future, a future formed by the divine tradition of the Astral Superintendent and the amicable dance of iridescent energies that characterized the city's presence.

2.3 Begin to unravel the rules and limitations of this unique ability.

In the rambling city of Etheria, where transcending high rises came to towards the sky and esoteric energies murmured in the air, carried on with a youthful individual named Lyra Shadowspire. Lyra, with her puncturing violet eyes and a fountain of 12 PM blue hair that appeared to retain the encompassing enchantment, had a one of a kind capacity that put her aside in a city where the limits between the ordinary and the esoteric obscured. Her gift, known as Aetherweaving, permitted her to control the actual texture of supernatural energies that penetrated Etheria.

Etheria, a city at the bleeding edge of hidden development, blossomed with the fragile harmony among innovation and sorcery. The residents, known as Etherians, had bridled the dormant esoteric energies to control their headways, from charmed transportation organizations to mysterious specialized gadgets. It was in this climate that Lyra found her abilities to aetherweaving, an ability that permitted her to wind around, shape, and control mystical energies with an artfulness that outperformed the capacities of her companions.

Lyra's excursion into understanding the guidelines and constraints of Aetherweaving started when, during a snapshot of extreme concentration, she invoked a sensitive radiant string of enchantment between her fingertips. This enchanted string, throbbing with ethereal energy, answered Lyra's aim, shaping many-sided designs as though directed by an undetectable hand. As she dove further into her recently discovered capacities, Lyra understood that Aetherweaving was not only a showcase of mysterious ability but rather a type of little known creativity.

In Etheria's eminent School of Little known Expressions, Lyra looked for direction from experienced mages and researchers who could reveal insight into the complexities of Aetherweaving. Among her guides was Teacher Arion Stormwind, a revered mage with a long, streaming facial hair that appeared to pop with static energy. Teacher Stormwind perceived the potential inside Lyra and encouraged her, directing her through the maze of supernatural hypothesis and functional applications.

As Lyra dove into the investigation of Aetherweaving, she revealed the key rules that administered the control of mysterious energies. The primary rule, as Teacher Stormwind clarified, was the idea of Reverberation — an association between the Aetherweaver's plan and the inborn idea of the mysterious energies. The reverberation went about as an impetus, permitting the Aetherweaver to shape and wind around the hidden flows as per their will.

Notwithstanding, with the force of Aetherweaving came the acknowledgment of impediments. Lyra found that the multifaceted nature of her weavings was straightforwardly relative to how she might interpret the supernatural energies at play. Complex spells and charms required a significant understanding of the esoteric powers, and the expectation to learn and adapt ended up being both requesting and fulfilling.

One more basic part of Aetherweaving that Lyra uncovered was the guideline of Equilibrium. The control of mystical energies, similar as a sensitive dance, required balance. Uncontrolled utilization of Aetherweaving could prompt otherworldly shakiness, causing eccentric peculiarities and possible kickback. Lyra figured out how to proceed cautiously, guaranteeing that her weavings kept an amicable harmony among goal and the normal progression of little known flows.

One of the most interesting parts of Aetherweaving that Lyra experienced was the thought of Supernatural Affinities. Each Aetherweaver, she found, had a one of a kind fondness towards particular sorts of mystical energies. Some succeeded in controlling essential powers, while others tracked down reverberation with recuperating or illusionary sorceries. Lyra, with her instinctive association with ethereal strings of sorcery, fostered a liking for winding around spells that impacted the texture of reality itself.

As Lyra's capability in Aetherweaving developed, so did the intricacy of her charms. She could wind around deceptions that deluded the faculties, make defensive hindrances that redirected esoteric assaults, and even control time for a little scope. However, with each newly discovered capacity, Lyra experienced the limits that compelled the maximum capacity of Aetherweaving.

The main impediment that surfaced was the idea of Supernatural Weakness. Aetherweaving requested a significant use of the Aetherweaver's mystical stores. Drawn out or excessively complex charms could prompt depletion, leaving the Aetherweaver depleted and defenseless. Lyra found that understanding her own mystical endurance and figuring out how to find a steady speed were crucial for excelling at Aetherweaving.

The subsequent limit uncovered itself through the multifaceted transaction of supernatural strings. Lyra found that specific enchanted weavings had unseen side-effects when consolidated. The Aetherweaver's purpose, on the off chance that not lined up with the inborn idea of the supernatural energies, could prompt unexpected outcomes. Lyra experienced minutes where the amicable reverberation she looked for was disturbed, bringing about spell failures to discharge or impermanent loss of command over the captivated strings.

In her journey to disentangle the guidelines of Aetherweaving, Lyra went up against moral contemplations. The ability to control reality and impact the personalities of others conveyed the heaviness of obligation.

Teacher Stormwind underscored the significance of moral Aetherweaving — involving the workmanship to improve society and regarding the independence of people. Lyra, presently mindful of the likely results of her activities, considered the ethical ramifications of using such impressive mysterious capacities.

As Lyra explored through the intricacies of Aetherweaving, Etheria confronted outside dangers that tried the city's otherworldly safeguards. A secrecy of maverick mages, outfitting illegal expressions, tried to weaken the fragile harmony among enchantment and innovation. The Etherian Board, perceiving the desperation of the circumstance, called upon talented Aetherweavers to invigorate the city's supernatural wards and safeguard against the looming obscure invasion.

Lyra, presently a perceived Aetherweaver with a developing standing, combined efforts with her individual mages to counter the maverick secrecy's maneuvers. The resulting supernatural fights unfurled in the core of Etheria, where the conflict of obscure energies enlightened the night sky. Lyra, drawing upon how she might interpret Aetherweaving, wove many-sided boundaries that repulsed the maverick mages' spells and upset their endeavors to penetrate the city's safeguards.

In the intensity of the supernatural struggle, Lyra experienced the genuine trial of her abilities to aetherweaving. The maverick mages, employing illegal charms that resisted regular figuring out, stretched the boundaries of Etheria's supernatural safeguards. Lyra, energized by the direness of the circumstance, dove into the undiscovered repositories of her mystical endurance, driving herself to wind around spells of exceptional intricacy.

As the fight arrived at its peak, Lyra confronted the moral situation of utilizing more intense charms that could have expansive outcomes. Teacher Stormwind's lessons reverberated to her, asking her to think about the drawn out impacts of her activities. In a mental breakthrough, Lyra decided to wind around a spell of limitation — an enchanted field that killed the maverick mages' capacities without truly hurting long-lasting.

The rebel secrecy, presently deprived of their esoteric powers, gave up to the Etherian Board. The city, because of the aggregate endeavors of talented Aetherweavers like Lyra, arose out of the contention sound. The occasions, in any case, developed's comprehension Lyra might interpret the moral obligations that went with the craft of Aetherweaving.

In the repercussions of the contention, Lyra proceeded with her examinations and investigation of Aetherweaving. The fights she battled and the difficulties she experienced became vital to her development as an Aetherweaver. She dove into the complexities of consolidating supernatural affinities, looking to make cooperative energies that enhanced her capacities without compromising the fragile equilibrium.

Lyra's excursion, from a fledgling Aetherweaver to a gifted expert, unfurled as a demonstration of the consistently unfurling intricacies of mystical control. The standards and constraints of Aetherweaving, once puzzling and cryptic, became strings woven into the texture of Lyra's mystical excursion. The city of Etheria, presently strengthened by the aggregate insight of its Aetherweavers, embraced a future where the sensitive dance among wizardry and innovation went on under the full concentrations eyes of the people who grasped the significant principles that represented the specialty of Aetherweaving.

In the otherworldly domain of Eldoria, where old woods murmured mysteries and ethereal creatures moved in the twilight knolls, carried on with a youthful individual named Elara Moonshadow. Elara, with her eyes mirroring the radiance of starlight and silver hair that flowed like moonbeams, had an uncommon and exceptional capacity known as Divine Tying. This capacity permitted her to fashion an association with the divine bodies that enhanced the night sky, directing their energies for different purposes. In any case, as Elara dug further into her heavenly gift, she started to unwind the multifaceted snare of limits that characterized the limits of this phenomenal power.

Eldoria, a domain saturated with wizardry and old legend, loved the divine bodies as heavenly divinities. The moon, the stars, and the heavenly bodies held significant importance, affecting the back and forth movement of wizardry that pervaded the land. Elara, brought into the world under a divine arrangement that noticeable her as a picked carrier of the Heavenly Tying, found her interesting association with the heavenly energies early on.

As Elara's capacities showed, so did the innate standards that represented Divine Tying. The most importantly limit that she experienced was the reliance on divine arrangements. The adequacy of Elara's capacities was unpredictably connected to the places of the moon, stars, and planets in the night sky. During heavenly occasions, for example, lunar stages or planetary arrangements, Elara's power flooded, permitting her to channel increased divine energies. Nonetheless, without any positive arrangements, her capacities were decreased, and getting to the full degree of Divine Tying became testing.

This impediment represented a critical test to Elara, particularly during divine peculiarities known as astral obscurations, where heavenly bodies briefly darkened one another. During such events, Elara's association with the divine energies wound down, leaving her briefly without admittance to the powers that characterized her personality. The recurrent idea of heavenly arrangements turned into a steady sign of the ephemerality of her capacities.

One more urgent part of Heavenly Tying that Elara found was the idea of Divine Congruity. The heavenly bodies worked in a fragile dance, each impacting the other-worldly flows in Eldoria. Elara's capacity to tackle divine energies was most strong when the heavenly bodies were in amicable arrangement.

Interruptions in this divine agreement, brought about by astronomical oddities or aggravations in the mystical equilibrium, could hinder Elara's capacity to successfully tie to the heavenly powers.

As Elara improved her abilities, she figured out how to adjust herself to the divine congruity, fostering a natural feeling of the back and forth movement of heavenly energies. Nonetheless, the consistently changing elements of Eldoria's enchanted scene implied that keeping up with this amicability required steady cautiousness and flexibility. Heavenly aggravations, unusual as they were, introduced difficulties that tried Elara's strength and genius.

A critical disclosure in Elara's investigation of Divine Tying was the idea of Heavenly Affinities. Each heavenly body, she understood, pervaded her with unmistakable powers and capacities. The moon, with its peaceful sparkle, truly Elara the capacity to improve mending and rebuilding. The stars, dispersed across the night sky like precious stones, presented to her increased instinct and foreknowledge. The planets, with their gravitational impact, empowered her to control the actual powers around her.

However, this divine abundance accompanied limits. Elara found that she could tie to each divine body in turn, picking the particular fondness that lined up with her goal. This impediment required cautious thought and vital preparation, particularly in snapshots of emergency when the decision of heavenly partiality could decide the result of a circumstance.

As Elara's standing as a Divine Tetherer developed, so did the assumptions set upon her by the occupants of Eldoria. They sought her during heavenly occasions for direction, trusting that her capacities could shield the domain from supernatural lopsided characteristics or divine unsettling influences. The strain to satisfy these assumptions became both an inspiration and a restriction for Elara, who wrestled with the heaviness of being a divine channel in a domain that loved the heavenly bodies.

The most significant restriction that Elara looked in her excursion of Divine Tying was the innate association between her feelings and the viability of her capacities. The heavenly energies answered not exclusively to the arrangement of divine bodies yet additionally to the profound reverberation inside Elara's heart. Positive feelings, like serenity and compassion, fortified her association with the divine powers, improving the strength of her capacities. Alternately, snapshots of close to home disturbance or pain could upset the sensitive equilibrium, prompting flighty results and lessening the adequacy of Heavenly Tying.

This profound reliance brought the two difficulties and disclosures for Elara. The contemplative idea of her capacities constrained her to stand up to and deal with her own feelings, changing her excursion into one of self-disclosure and close to home strength. Heavenly Tying, it appeared, was a supernatural association with the universe as well as a mirror mirroring the deepest sensations of its conveyor.

In the midst of individual unrest, Elara wrestled with the battle to keep up with command over her capacities. The divine energies, receptive to the variances of her feelings, answered with both compassion and capriciousness. It was during these minutes that Elara looked for comfort in the insight of Eldoria's otherworldly substances and looked for direction from the divine creatures she adored.

In a snapshot of disclosure, Elara figured out how to explore the many-sided dance between her feelings and Divine Tying. She embraced the weakness of her feelings, perceiving that the divine energies answered not to a veneer of emotionlessness but rather to the legitimacy of her sentiments. This freshly discovered understanding permitted her to fashion a more significant association with the divine powers, changing snapshots of inner disturbance into open doors for development and strengthening.

As Elara developed in her job as a Divine Tetherer, she turned into an epitome of the fragile harmony between the heavenly and the human domains. Her process unfurled as an embroidery woven with divine strings, every constraint and disclosure adding to the rich story of Eldoria's mysterious woven artwork.

In the core of Eldoria, where Divine Tying held a strange power, Elara Moonshadow's presence turned into an image of strength, flexibility, and the significant association between the heavenly and the human domains. The impediments inborn in her novel capacity became strings that fortified the texture of her personality, and the divine bodies above gave testimony regarding the developing adventure of a youthful person who embraced both the enormous and the human components of her reality.

Chapter 3

Degrees of Despair

In the domain of human experience, there exists a range of feelings that cross the huge scene of the mind. Among these, despair remains as a significant and tormenting presence, creating its for some time shaded areas across the hallways of the human spirit. Levels of sadness fluctuate, going from the unpretentious reverberations of despairing to the staggering chasm of complete sadness.

At the shallow finish of this profound pool lies a delicate despairing, an insightful bitterness that sticks to the edges of our cognizance like morning fog. It's the tranquil throb of unfulfilled dreams and the delicate moan of yearnings that have lost their gloss. This unpretentious kind of gloom frequently slips by everyone's notice, covered by the back and forth movement of regular day to day existence. The despairing goes with a blustery day or the blurring sparkle of a dusk, abandoning a feeling of transient excellence that can never be completely gotten a handle on.

As we venture further into the openings of gloom, we experience the more discernible load of dissatisfaction. This is a misery brought into the world from the hole among assumption and reality, a harsh taste that waits on the tongue of the people who hoped against hope. Bombed connections, unfulfilled aspirations, and the sting of disloyalty all add to this more articulated type of sadness. It's the premonition in the pit of the stomach when life goes off in a strange direction, and the street ahead is clouded by an impervious obscurity of vulnerability.

Moving further along the range, we end up in the domain of sorrow. Here, the close to home scene is described by a feeling of weakness and an unavoidable sensation of being caught in an unpreventable trap of situation. Dejection is the buddy of the people who have confronted rehashed misfortunes and found their strength worn ragged. It's the acknowledgment that the walls shutting in are through our own effort, developed one step at a time from the decisions and activities that misled us.

However, even in the most obscure corners of sadness, there exists a profundity that rises above the simple shortfall of trust. Sheer sadness is a gorge so significant that it swallows the actual idea of a future, leaving just the hopeless breadth of the current second. This is the territory of the clinically discouraged, where the psyche turns into a landmark, and each step in the right direction is an exhausting excursion through the entanglement of a dreary presence.

The levels of despondency are not static; they stream and interlace, making an intricate embroidery of human experience. Similarly as satisfaction can be viewed as in the littlest of triumphs, gloom can be brought into the world from the collection of apparently irrelevant losses. It's a fragile dance among light and shadow, worked out on the phase of our lives.

In investigating the profundities of gloom, it becomes obvious that the human experience is set apart by a significant interconnectedness. The profound flows that go through every one of us resemble feeders taking care of into an extraordinary stream of shared cognizance. At the point when one individual wrestles with despair, the waves reach out a long ways past the limits of their own battles, contacting the existences of everyone around them.

Consider the instance of Emily, a lady whose life appeared to be a material painted with the energetic tints of progress and satisfaction. She had a prospering profession, a caring family, and a friend network who respected her strength. However, underneath the surface, a situation was unfolding. Emily's inward scene was defaced by the quiet disintegration of her psychological prosperity.

As the tension built, Emily wound up trapped in the undertow of hopelessness. The heaviness of assumptions, both outside and self inflicted, turned into an anchor hauling her more profound into the dim waters of devastation. Each grin she wore was a veil covering the breaks inside, and each accomplishment was a short lived float in the whirlwind of her unwinding mental state.

The levels of despondency appeared in Emily's life were not disconnected; they sent shockwaves through the complicated trap of connections she had developed. Her family, when floated by the glow of her presence, presently wound up exploring the capricious flows of her inner disturbance. Companions, acquainted with depending on Emily for help, were presently confronted with a perturbing job inversion.

As Emily's despondency extended, it turned into a common encounter, an immaterial string associating divergent lives in a sensitive dance of aggregate torment. The levels of despondency, once bound to the individual, presently reverberated through the spaces between individuals, winding around a story of shared misery.

In the more extensive setting of society, the levels of gloom take on an aggregate shade. Social, monetary, and political variables add to the back and forth movement of hopelessness on an excellent scale. In people group minimized by foundational disparity, hopelessness can turn into an unavoidable power, molding the shared perspective and impacting the direction of whole populaces.

Consider a cityscape set apart by haggard structures, where the reverberations of despondency resonate through the roads like an eerie tune. Here, the levels of sadness are scratched into the actual texture of the local area, appearing in the essences of the people who explore a scene without any trace of chance. The pattern of destitution, filled by foundational treacheries, turns into a favorable place for despair, sustaining a heritage that stretches across ages.

In such conditions, the levels of hopelessness are an individual battle as well as an impression of the more extensive cultural discomfort. The individual and the aggregate become inseparably connected, each impacting and building up the other in a criticism circle of despondency. Breaking liberated from this cycle requires individual versatility as well as an aggregate work to destroy the designs that sustain imbalance and cutoff the opportunities for a superior future.

The levels of depression are not bound to the human experience alone; they broaden their rings into the regular world. Ecological debasement, loss of biodiversity, and the approaching phantom of environmental change cast a shadow that contacts each living being on this planet. The hopelessness brought into the world from the debasement of the climate is certainly not a far off concern however a substantial reality that saturates the very air we inhale and the water we drink.

As the outcomes of human activities in the world become progressively evident, the levels of gloom heighten. The biting tension goes with the acknowledgment of an irreversible misfortune, the destruction of seeing environments disintegrate, and the vulnerability notwithstanding a planet pushed as far as possible. The ecological emergency is an aggregate test that requests an aggregate reaction, rising above boundaries and belief systems chasing a reasonable future.

In the embroidery of depression, there are likewise strings of flexibility that weave a counter-story. Mankind's set of experiences is loaded with accounts of people and networks who, despite outlandish chances, have ascended from the profundities of hopelessness to fashion a way of recovery. These accounts are a demonstration of the dauntless soul that lives inside us, equipped for rising above even the haziest shades of depression.

Think about the tale of Maria, an overcomer of war and uprooting who wound up in an exile camp, encompassed by the flotsam and jetsam of broken lives. The levels of gloom in Maria's life were significant, as she wrestled with the injury of misfortune and the vulnerability of an obscure future. However, inside the limits of that displaced person camp, Maria found a wellspring of versatility that challenged the severe load of sadness.

Amidst misfortune, Maria turned into an encouraging sign for people around her. She coordinated improvised schools for youngsters, offering a glint of business as usual amidst mayhem. Her flexibility turned into an impetus for aggregate activity, rousing others to channel their despondency into a power for positive change. Maria's story is definitely not a separated occurrence yet an impression of the groundbreaking power implanted inside the human soul.

The levels of sadness, when met with versatility, can possibly catalyze significant individual and cultural change. It's in the cauldron of hopelessness that people find the dormant strength inside, a strength that empowers them to stand up to the shadows and arise with recently discovered reason. In this catalytic cycle, despair turns into a cauldron for versatility, producing people who are not characterized by their enduring yet engaged by their capacity to transcend it.

As we explore the maze of gloom, it becomes obvious that the excursion isn't straight. It's a wandering way set apart by exciting bends in the road, where snapshots of brightening are frequently gone before by stretches of significant haziness. The levels of sadness are not fixed; they advance and adjust, answering the moving flows of life.

In the dance among misery and flexibility, there exists a sensitive equilibrium that requires both reflection and aggregate activity. Individual change, however fundamental, isn't adequate in disengagement. The levels of depression that pervade society request an aggregate retribution — a common obligation to destroy the designs that sustain disparity, unfairness, and ecological corruption.

To stand up to the levels of depression on a cultural level is to recognize the interconnectedness of our destinies. It requires an acknowledgment that the prosperity of people is indivisible from the prosperity of networks, countries, and the planet all in all. In this acknowledgment lies the seed of trust — an expectation that, when sustained all in all, has the ability to break the shackles of gloom and usher in another time of plausibility.

All in all, the levels of sadness structure a mind boggling embroidery woven from the strings of individual and aggregate encounters. From the unpretentious despairing that colors our day to day routines to the significant chasm of sheer sadness, despair appears in horde structures, forming the shapes of the human excursion. A power can both segregate and interface, creating shaded areas that stretch across the scenes of our own and cultural stories.

However, inside the profundities of misery, there exists a potential for change. The levels of depression, when met with strength and aggregate activity, can turn into an impetus for positive change. The human soul, versatile and dauntless, has the ability to transcend the shadows and arise into the illumination of another day.

As we explore the levels of hopelessness in our own lives and in the more extensive setting of the world, let us recollect that we are in good company. The embroidery of despondency is woven with the strings of shared mankind, and in that common experience, there is a hint of something to look forward to. In the acknowledgment of our interconnectedness lies the chance of a future where the levels of hopelessness are tempered by the strength of the human soul and the aggregate will to construct a superior world.

3.1 Present a crisis in the world related to extreme temperatures.

In the consistently changing embroidery of our reality, the ghost of environmental change looms as a significant and critical emergency, with outrageous temperatures arising as perhaps of its most substantial and damaging sign. The World's environment is a fragile harmony between interconnected frameworks, and disturbances to this balance have extensive results. As worldwide temperatures climb at a phenomenal rate, the world wrestles with the repercussions of outrageous intensity, leading to an emergency that contacts each side of the globe.

One of the most obvious indications of the strengthening environment emergency is the expansion in the recurrence and seriousness of heatwaves. Across mainlands, from North America to Europe, Asia to Africa, heatwaves have become something other than meteorological peculiarities; they are presently harbingers of a planetary irregularity that jeopardizes environments, economies, and the prosperity of billions of individuals.

As of late, record-breaking temperatures have become unnervingly typical. Urban communities once acclimated with moderate environments currently end up boiling under the abusive intensity of drawn out heatwaves. The results of outrageous temperatures stretch out past simple uneasiness; they penetrate each part of human existence, from agribusiness to general wellbeing, from energy utilization to social elements.

Farming, the bedrock of human civilization, bears a significant weight even with outrageous temperatures. Rising intensity represents an immediate danger to edit yields and food security. Staple harvests, like wheat, rice, and corn, face decreased yields and lower dietary benefit under states of outrageous intensity.

Drawn out heatwaves fuel dry season conditions, prompting water shortage and further pushing farming frameworks previously stressed by the evolving environment.

In districts vigorously reliant upon horticulture, the effect of outrageous temperatures resounds through country networks. Ranchers wrestle with bombing crops, reducing animals efficiency, and the apparition of financial breakdown. The emergency reaches out past individual jobs; it upsets worldwide food supply chains, raising the phantom of food deficiencies and cost spikes that lopsidedly influence weak populaces.

Past the domain of horticulture, outrageous temperatures represent an immediate danger to human wellbeing. Heatwaves carry with them a large group of wellbeing gambles, going from heat pressure and lack of hydration to additional extreme circumstances like heatstroke. Weak populaces, including the old, youngsters, and those with prior medical issue, face increased takes a chance during episodes of outrageous intensity.

Metropolitan regions, with their substantial wildernesses and intensity engrossing framework, become authentic intensity islands during heatwaves. The constructed climate traps and enhances heat, making pockets of intolerable temperatures. The outcomes are desperate, with expanded paces of intensity related sicknesses and a flood popular for crisis clinical benefits. The actual texture of metropolitan life is stressed as urban communities wrestle with the requirement for cooling focuses, crisis reaction plans, and long haul techniques to adjust to the new typical of climbing temperatures.

The energy scene, as well, goes through a change notwithstanding outrageous temperatures. The interest for cooling, driven by the need to get away from the intense intensity, puts massive weight on energy lattices. Cooling units, when an extravagance, become a life saver during heatwaves, driving up power utilization to unreasonable levels. The subsequent stress on energy framework raises worries about unwavering quality as well as fuels the carbon impression, adding to an endless loop of environmental change.

The effect of outrageous temperatures stretches out past human boundaries, influencing environments and biodiversity on a worldwide scale. From coral reefs dying under the searing sun to polar ice covers liquefying at a disturbing rate, the normal world is a quiet survivor of the rising intensity. Species face annihilation as they battle to adjust to evolving environments, and fragile biological systems, finely tuned to explicit temperature ranges, unwind notwithstanding extraordinary intensity.

In the Cold, the fast warming has significant ramifications for native networks whose lifestyle is complicatedly attached to the ice-shrouded scenes. Conventional practices, for example, hunting and fishing become dangerous as ice retreats, disturbing age-old rhythms and testing the flexibility of societies adjusted to the brutal Icy circumstances.

The emergency of outrageous temperatures isn't restricted to explicit locales; a worldwide test requests aggregate activity and a reexamination of humankind's relationship with the planet.

The results of outrageous temperatures are not disseminated evenhandedly, further highlighting the social components of the emergency. Minimized people group, frequently situated in regions helpless against environment influences, bear a lopsided weight. The convergence of environmental change and social disparities intensifies existing weaknesses, passing on underestimated populaces with less assets to adjust and recuperate from the effects of outrageous intensity.

In non-industrial countries, where admittance to assets and foundation is restricted, the emergency of outrageous temperatures develops existing difficulties. Deficient lodging, absence of admittance to clean water, and restricted medical care assets compound the dangers looked by networks currently on the edges. The worldwide idea of environmental change requests a worldwide reaction, one that recognizes the interconnectedness of natural and civil rights.

The emergency of outrageous temperatures is certainly not a far off danger; a reality requests quick and purposeful activity. The Paris Understanding, a milestone worldwide accord, frames a system for worldwide endeavors to moderate environmental change and adjust to its effects. Notwithstanding, the earnestness of the circumstance requires something beyond responsibilities on paper; it requires an extreme change in the manner in which social orders see and answer the environment emergency.

Alleviation endeavors should zero in on decreasing ozone depleting substance outflows, the essential driver of environmental change. Progressing to environmentally friendly power sources, upgrading energy proficiency, and reconsidering metropolitan wanting to focus on maintainability are basic parts of an extensive relief procedure. The shift to a low-carbon economy isn't just an ecological objective yet in addition a monetary open door that can drive development, make occupations, and encourage versatile networks.

Transformation is similarly vital notwithstanding the unavoidable effects of environmental change. Putting resources into tough framework, executing maintainable farming practices, and growing early admonition frameworks for outrageous climate occasions are fundamental parts of variation methodologies. The objective isn't simply to endure the hardship yet to fabricate social orders that can flourish despite an evolving environment.

Worldwide collaboration is vital in tending to the emergency of outrageous temperatures. Environmental change knows no lines, and its effects are felt worldwide. Created countries bear an obligation to help emerging nations in their endeavors to adjust and relieve, perceiving the verifiable commitments to ozone harming substance outflows and the differential abilities to answer the emergency.

Individual activity, as well, assumes an essential part in the aggregate reaction to the emergency. Maintainable living practices, cognizant utilization, and backing for environment well disposed strategies add to the more extensive development for natural stewardship. The emergency of outrageous temperatures is a source of inspiration that resounds at the individual, local area, and cultural levels.

All in all, the emergency of outrageous temperatures remains as a characterizing challenge within recent memory, one that requests a thorough and facilitated reaction. The effects of climbing temperatures contact each aspect of human existence, from the food we eat to the air we relax. The interconnectedness of the emergency highlights the requirement for a comprehensive methodology that tends to the underlying drivers of environmental change, focuses on variation and versatility, and perceives the social elements of the test.

As the world wrestles with the ramifications of outrageous temperatures, the decisions before very long will shape the direction of our common future. The emergency isn't unrealistic, yet it requires an aggregate obligation to head in a different direction, rethink needs, and produce a way towards a maintainable and versatile world. The criticalness existing apart from everything else is an unmistakable update that the ideal opportunity for activity is presently, and the decisions today will reverberate for a long time into the future.

3.2 The protagonist faces their first major challenge, showcasing the potential consequences of misusing their power.

In the domain of narrating, the hero's process is frequently set apart by hardships, filling in as a cauldron for self-improvement and advancement. It is inside these difficulties that the real essence of the hero's personality is uncovered, and their capacity to explore misfortune is scrutinized. In this story, the hero faces their most memorable significant test, a crucial second that shapes their fate as well as features the expected results of abusing the unprecedented power they have.

The hero, a youthful individual enriched with an interesting and impressive influence, sets out on an excursion that rises above the limits of the normal. This power, whether otherworldly, mechanical, or inborn, separates them from the world they occupy. In the underlying phases of their excursion, the hero is directed by a feeling of miracle and revelation, anxious to investigate the boundless conceivable outcomes that their power manages the cost of them.

As the story unfurls, the hero experiences a general public wrestling with its own arrangement of difficulties and clashes. The world they possess isn't insusceptible to the intricacies of force elements, cultural designs, and moral quandaries. The hero's power, at first a wellspring of freedom and fervor, turns into a situation with two sides as they explore the complicated snare of connections and obligations that characterize their reality.

The main significant test arises as an outcome of the hero's activities, a far reaching influence of their decisions with their recently discovered power. It starts honestly enough — an exhibit of their capacities to a limited handful, a longing to dazzle or show what them can do. In any case, the results heighten suddenly, and the hero ends up ensnared in a trap of accidental results.

The abuse of force, whether driven by presumption, obliviousness, or a blend of both, gets rolling a chain of occasions that resonates through the story. The once clear qualifications among good and bad haze, and the hero is stood up to with the results of their activities. The very power that was intended to engage turns into a wellspring of strife, creating a shaded area over the hero's excursion.

One part of the test lies in the cultural reaction to the hero's abuse of force. The people group, at first captivated by the hero's capacities, moves its discernment as the outcomes unfurl. Trust is disintegrated, and incredulity gives method for dreading and doubt. The hero, when viewed as an encouraging sign, presently turns into an image of the potential risks related with unrestrained power.

The far reaching influence stretches out to the connections the hero holds dear. Companions, guides, and partners who once remained close by now question the moral limits of the power the hero employs. Selling out and dissatisfaction variety these connections, as those nearest to the hero wrestle with the consequences of their activities. The test becomes a trial of the hero's personality as well as a pot for the bonds they have fashioned.

Inside, the hero wrestles with a significant feeling of culpability and obligation. The outcomes of their abuse of force weigh vigorously on their soul, provoking a profound thoughtfulness into the idea of ethical quality, responsibility, and the genuine importance of the capacities they have. The conflict under the surface turns into a characterizing part of the test, as the hero stands up to the hazier parts of their own longings and expectations.

The outer and inside components of the test join as the hero faces an ethical situation. A critical decision looms not too far off, requesting a choice that will shape the direction of their excursion. Recovery and the potential for development coax on one way, while the charm of additional abuse of force and the quest for individual increase captivate on the other. The hero remains at an intersection, a snapshot of significant importance that typifies the pith of their personality circular segment.

In this pot of decisions, the hero's choice conveys sweeping results. The story unfurls along dissimilar ways, each prompting an unmistakable arrangement of results. The repercussions of the hero's decision influence their own process as well as send swells through the world they possess. The test turns into a defining moment, a snapshot of retribution that shapes the hero's character and establishes the vibe for the story's unfurling sections.

In the event that the hero picks the way of reclamation and development, the story turns into an investigation of strength, self-revelation, and the limit with regards to positive change. The hero wrestles with the aftermath of their prior activities, looking to fix cracked connections, revamp trust, and saddle their power for everyone's best interests. The test turns into an impetus for change, and the hero rises out of the pot with freshly discovered intelligence and a more profound comprehension of the obligations that go with their capacities.

On the other hand, on the off chance that the hero capitulates to the charm of proceeded with abuse of force, the story takes a more obscure turn. The results of their decisions raise, and the hero becomes entrapped in a snare of double dealing, disloyalty, and the disintegration of their own mankind. The test, rather than filling in as a snapshot of development, turns into a plunge into moral uncertainty, power-driven pride, and the likely distance of the hero from the people who once remained close by.

Regardless, the story winds around an embroidery of results that stretches out past the hero's quick circle. The cultural reaction, the effect on connections, and the hero's subtle conflicts resound through the all-encompassing account, impacting the direction of the story. The test, however a solitary occasion, turns into a topical string that ties together the different components of the hero's excursion.

As the story unfurls, the hero's reaction to the main significant test turns into a mirror mirroring the intricacies of human instinct and the nuanced investigation of force elements. The story turns into a critique on the delicacy of profound quality, the enticements of uncontrolled power, and the redemptive likely intrinsic in the human soul.

All through the hero's excursion, the principal significant test fills in as a standard, a reference point that reverberations through resulting preliminaries and quandaries. The illustrations realized, whether through the hug of development and recovery or the plunge into dimness, shape the hero's advancing person. The outcomes of abusing power become a common theme, a useful example that impacts the decisions the hero makes as they explore the exciting bends in the road of their story circular segment.

All in all, the principal significant test looked by the hero turns into an essential second in the unfurling story. A pot tests the hero's personality, exhibiting the possible outcomes of abusing their power. The test reaches out past the quick decisions and activities, turning into a topical string that winds through the general story. The hero's reaction to this challenge turns into a focal point through which the intricacies of human instinct, power elements, and the quest for reclamation are investigated. As the story unfurls, the repercussions of the hero's decisions act as a directing power, forming their excursion and impacting the direction of the story.

3.3 Develop interpersonal relationships as the protagonist seeks alliances to overcome the crisis.

In the complex embroidery of narrating, relational connections act as a crucial string that winds around together the texture of a person's excursion. The hero, exploring the maze of difficulties and emergencies, leaves on a journey to produce unions that reach out past the limits of individual capacities. These connections become the key part in the hero's interest to beat the approaching emergency, a demonstration of the force of association and coordinated effort.

As the story unfurls, the hero experiences a different exhibit of characters, each with their own inspirations, foundations, and abilities. These characters become expected partners, people whose qualities supplement the hero's own and whose remarkable points of view add to an aggregate comprehension of the emergency within reach. The most common way of creating relational connections turns into a nuanced dance, requiring compassion, trust-building, and a common feeling of direction.

The main seeds of union are much of the time established in the prolific ground of shared objectives and common interests. The hero, driven by the direness of the emergency, searches out people whose goals line up with their own. Whether roused by a common feeling of obligation, a shared adversary, or an aggregate vision for a superior future, these unions structure the establishment whereupon the hero's process is fabricated.

Chasing collusions, the hero participates in a sensitive equilibrium between weakness and strength. Opening oneself to the complexities of relational associations requires an eagerness to uncover one's own weaknesses, fears, and instabilities. In this weakness trust is developed, the bedrock whereupon getting through unions are built. The hero, in spite of the weights they convey, figures out how to impart the heaviness of their excursion to other people, encouraging a feeling of kinship and shared liability.

The improvement of relational connections is definitely not a straight interaction yet a dynamic and developing excursion. The hero explores the recurring patterns of association, confronting snapshots of pressure, conflict, and misconstruing. These difficulties become open doors for development, as the connections between partners are tried and reinforced through shared difficulty. The extravagance of relational connections lies not in their consistent flawlessness but rather in their ability to weather conditions storms and arise strong.

One critical part of union structure is the trading of information and abilities. Each partner brings an extraordinary arrangement of capacities, encounters, and points of view to the table. The hero, perceiving the worth of different skill, takes part in an equal trade of bits of knowledge and capacities. This cooperative learning turns into a competitive edge, upgrading the aggregate ability to explore the intricacies of the emergency.

In the embroidery of relational connections, mentorship arises as a significant string. The hero, frequently directed by experienced and prepared partners, benefits from the insight and direction granted by the individuals who have track comparable ways. Mentorship turns into a wellspring of solidarity, offering the hero viable information as well as a feeling of consolation and an update that they are in good company in their excursion.

Companionship, as well, blooms inside the cauldron of shared difficulties. Partners become compatriots, people with whom the hero can share snapshots of delight, distress, and in the middle between. The close to home reverberation of these connections adds profundity to the story, establishing the hero in an organization of help that reaches out past the items of common sense of the emergency. Companionship turns into a wellspring of comfort and consolation, a sign of the human associations that persevere through even notwithstanding misfortune.

Nonetheless, not all relational connections are produced in the pot of fellowship. The hero, as they continued looking for collusions, experiences people with clashing plans, stowed away thought processes, and disparate viewpoints. These hostile connections acquaint a layer of intricacy with the story, driving the hero to explore the sensitive harmony among collaboration and dispute.

The test of overseeing clashing collusions turns into a trial of the hero's strategic abilities. Exploring the mind boggling snare of relational elements requires a sharp comprehension of the inspirations and desires of each partner. The hero turns into a go between, a scaffold developer who looks for shared belief even in the midst of unique targets. This political artfulness turns into a significant resource chasing a unified front against the emergency.

As the hero develops their unions, a local area starts to arise — an aggregate power more prominent than the amount of its singular parts. The people group turns into a microcosm of shared values, different gifts, and common help. It is inside this local area that the hero tracks down strength, versatility, and a feeling of having a place. The connections created all through the story mix into an organization of interconnected lives, each adding to the overall mission to conquer the emergency.

The relational connections produced by the hero reach out past the prompt circle of partners to incorporate more extensive organizations and coalitions. The hero, perceiving the interconnectedness of their reality, looks to construct spans with networks, associations, and even adversary groups. The partnerships become a web that ranges geological, social, and philosophical limits, mirroring the's comprehension hero might interpret the emergency as a worldwide test that requests a cooperative reaction.

Chasing more extensive partnerships, the hero turns into a representative on the world stage. Exchanges, settlements, and key associations become devices in the munititions stockpile, and the hero explores the intricacies of worldwide relations.

The stakes are higher, the results more significant, and the collusions more sensitive as the hero endeavors to fabricate an alliance equipped for tending to the emergency on a worldwide scale.

However, the improvement of relational connections isn't exclusively a necessary evil yet an inborn part of the hero's self-awareness. Through the associations produced with partners, coaches, and companions, the hero goes through a change — an excursion from detachment to interconnectedness, from independence to coordinated effort. The story turns into a demonstration of the groundbreaking force of human association, delineating how the hero's capacity to defeat the emergency is unpredictably attached to their ability to construct significant connections.

In the climactic snapshots of the account, the strength of the partnerships is put to a definitive test. The emergency, once approaching not too far off, presently unfurls in the entirety of its power. The hero, encompassed by partners and reinforced by the connections developed all through their excursion, faces the finish of their journey. The coalitions become a safeguard against the surge of difficulties, an aggregate power that stands against the tide of misfortune.

The goal of the emergency isn't exclusively dependent upon the hero's singular ability however on the cooperative endeavors of the whole organization of partners. Every individual from the local area, every collusion produced, adds to the aggregate strength and inventiveness expected to beat the emergency. The hero, when a single figure wrestling with the heaviness of their power, presently remains at the front of a unified front — an encouraging sign notwithstanding vulnerability.

The repercussions of the emergency turns into an impression of the strength of the relational connections created all through the story. The people group rises out of the cauldron of difficulties more grounded, more associated, and strong. The partnerships, once tried by difficulty, demonstrate their persevering through esteem as the heroes and their partners revamp, recuperate, and graph a course towards a fresh start.

All in all, the story of the hero's journey to defeat an emergency is complicatedly woven with the strings of relational connections. The partnerships produced all through the excursion become a main impetus, forming the hero's capacity to explore difficulties, beat misfortune, and fabricate a local area of mutual perspective. The advancement of connections, whether brought into the world of shared objectives, mentorship, fellowship, or strategic artfulness, adds profundity to the story, showing the extraordinary force of human association. In the embroidery of narrating, relational connections stand as a demonstration of the hero's ability to rise above individual constraints and face the approaching emergency with a unified front.

In the unfurling story of conquering an emergency, the job of partnerships arises as a key part in the hero's excursion. As the emergency looms not too far off, the hero explores a perplexing trap of connections, trying to fashion collusions that reach out past the limits of individual capacities. These coalitions become an essential need as well as a demonstration of the's comprehension hero might interpret the force of cooperation despite difficulty.

The excursion starts with the hero, remaining at the slope of an emergency, perceiving the intrinsic difficulties that lie ahead. The extent of the looming danger requires a shift from individual heroics to a more aggregate and cooperative methodology. In this acknowledgment, the hero sets out on a journey to frame partnerships, understanding that the qualities of many are definitely more imposing than the capacities of one.

Collusions, in their beginning phases, frequently sprout from the rich ground of shared goals and common interests. The hero, driven by the earnestness of the emergency, searches out people whose objectives line up with their own. Whether spurred by a common feeling of obligation, a shared adversary, or an aggregate vision for a superior future, these incipient collusions become the establishment whereupon the hero's process is fabricated.

The improvement of coalitions is certainly not a simple value-based trade yet a nuanced dance of relational elements. The hero, in their interest, learns the fragile specialty of trust-building, sympathy, and mutual perspective. The weakness expected to open oneself to the complexities of relational associations turns into a pivotal part of this union structure process. Trust, the bedrock whereupon persevering through partnerships are developed, is developed through shared encounters, common regard, and a common obligation to beating the emergency.

As the hero manufactures further associations, the story unfurls with a rich embroidery of characters, each contributing their extraordinary assets, foundations, and points of view to the aggregate exertion. These partners become mainstays of help, offering substantial abilities and assets as well as different perspectives that enhance the's comprehension hero might interpret the emergency and expected arrangements.

The most common way of creating relational connections inside partnerships is certainly not a direct movement however a dynamic and developing excursion. The hero explores the rhythmic movements of association, confronting snapshots of strain, conflict, and misjudging. These difficulties become open doors for development, as the connections between partners are tried and reinforced through shared affliction. The extravagance of relational connections lies not in their consistent flawlessness but rather in their ability to weather conditions storms and arise versatile.

One huge part of collusion building lies in the trading of information and abilities. Each partner offers an interesting arrangement of capacities that would be useful, and the hero perceives the worth of different skill.

The cooperative discovering that follows turns into a competitive edge, improving the aggregate ability to explore the intricacies of the emergency. This common information isn't just a competitive edge yet additionally a demonstration of the synergistic possible intrinsic in collusions.

In the embroidery of relational connections, mentorship arises as a critical string. The hero, frequently directed by experienced and prepared partners, benefits from the insight and direction bestowed by the people who have track comparable ways. Mentorship turns into a wellspring of solidarity, offering useful information as well as a feeling of consolation and an update that they are in good company in their excursion.

Companionship, as well, blooms inside the pot of shared difficulties. Partners become comrades, people with whom the hero can share snapshots of delight, distress, and in the middle between. The close to home reverberation of these connections adds profundity to the story, establishing the hero in an organization of help that reaches out past the items of common sense of the emergency. Kinship turns into a wellspring of comfort and consolation, a sign of the human associations that persevere through even notwithstanding misfortune.

In any case, not all relational connections inside unions are portrayed by kinship. The hero, as they continued looking for coalitions, experiences people with clashing plans, stowed away thought processes, and unique viewpoints. These opposing connections acquaint a layer of intricacy with the story, constraining the hero to explore the fragile harmony among collaboration and dispute.

The test of overseeing clashing unions turns into a trial of the hero's discretionary abilities. Exploring the multifaceted snare of relational elements requires a sharp comprehension of the inspirations and desires of each partner. The hero turns into a middle person, a scaffold developer who looks for shared view even in the midst of dissimilar goals. This discretionary artfulness turns into an important resource chasing a unified front against the emergency.

As the hero extends their partnerships, a local area starts to arise — an aggregate power more prominent than the amount of its singular parts. The people group turns into a microcosm of shared values, various gifts, and common help. It is inside this local area that the hero tracks down strength, flexibility, and a feeling of having a place. The connections created all through the story combine into an organization of interconnected lives, each adding to the overall journey to conquer the emergency.

The improvement of relational connections isn't exclusively a necessary evil yet a characteristic part of the hero's self-awareness. Through the associations manufactured with partners, guides, and companions, the hero goes through a change — an excursion from disengagement to interconnectedness, from independence to joint effort.

The story turns into a demonstration of the groundbreaking force of human association, delineating how the hero's capacity to conquer the emergency is complicatedly attached to their ability to construct significant connections.

In the climactic snapshots of the account, the strength of the collusions is put to a definitive test. The emergency, once approaching not too far off, presently unfurls in the entirety of its force. The hero, encompassed by partners and supported by the connections developed all through their excursion, faces the zenith of their mission. The unions become a safeguard against the invasion of difficulties, an aggregate power that stands against the tide of difficulty.

The goal of the emergency isn't exclusively dependent upon the hero's singular ability however on the cooperative endeavors of the whole organization of partners. Every individual from the local area, every union fashioned, adds to the aggregate strength and resourcefulness expected to beat the emergency. The hero, when a singular figure wrestling with the heaviness of their power, presently remains at the very front of a unified front — an encouraging sign notwithstanding vulnerability.

The consequence of the emergency turns into an impression of the strength of the relational connections created all through the story. The people group rises up out of the cauldron of difficulties more grounded, more associated, and versatile. The coalitions, once tried by misfortune, demonstrate their persevering through esteem as the heroes and their partners modify, recuperate, and outline a course towards a fresh start.

All in all, the story of the hero's mission to conquer an emergency is unpredictably woven with the strings of relational connections. The collusions manufactured all through the excursion become a main impetus, forming the hero's capacity to explore difficulties, defeat difficulty, and construct a local area of common perspective. The advancement of connections, whether brought into the world of shared objectives, mentorship, kinship, or discretionary artfulness, adds profundity to the story, delineating the groundbreaking force of human association. In the embroidery of narrating, relational connections stand as a demonstration of the hero's ability to rise above individual limits and face the approaching emergency with a unified front.

Chapter 4

Fahrenheit Friends

Not long from now, a world grasped by the persistent hug of innovation and social congruity unfurled. Here the temperature of human connections was misleadingly controlled, not by the glow of certified associations, but rather by the cool accuracy of a cultural indoor regulator. It was a tragic reality where close to home bonds were exposed to a Fahrenheit size of kinship, and the glow of brotherhood was estimated with clinical separation.

In this world, kinships were classified into severe mathematical qualities, each comparing to a particular Fahrenheit degree. The higher the number, the hotter and more significant the fellowship was considered to be. Individuals were assessed in view of their capacity to keep up with companionships inside this foreordained scale, and cultural assumptions directed the OK reach for a singular's Fahrenheit Companions score.

Our hero, Alex, explored this profoundly controlled social scene with a feeling of separation. He had consistently felt a discord between the fake warmth forced by cultural standards and the certified associations he hungered for. In spite of the consistent strain to adjust, Alex stayed unfaltering as he continued looking for bona fide connections, determined by the mathematical limitations that administered his general surroundings.

As Alex traveled through the carefully arranged schedules of day to day existence, he experienced people who tested the laid out request. These agitators, named as "Anomalies," resisted the standards and considered fashioning companionships that rose above the endorsed Fahrenheit cutoff points. The Exceptions resided on the edges of society, existing in a domain where the glow of human association was not limited by mathematical limitations.

One such Anomaly, Mia, entered Alex's life startlingly. Mia was a nonconformist who esteemed the profundity of profound association over cultural standards. She saw past the Fahrenheit Companions score, perceiving the impediments forced on connections by a framework that decreased human associations with simple digits. Mia's proud way to deal with kinship ignited an adjustment of Alex, provoking him to scrutinize the unbending designs that represented his reality.

Together, Alex and Mia set out on an excursion to find the genuine importance of fellowship in a general public that tried to measure and control it. As they dug further into the secrets of human association, they experienced other people who shared their longing for genuineness. The gathering of revolutionaries developed, every part contributing a special point of view to the developing story of opposition against the Fahrenheit Companions system.

As the renegades rocked the boat, they confronted resistance from the authorities of similarity. The Fahrenheit Companions Department, a strong substance that kept up with command over the social texture, tried to stifle any contradiction that compromised the laid out request. The radicals, nonetheless, stayed unflinching, energized by the conviction that veritable associations couldn't be kept to the restricted limits of a mathematical scale.

The story unfurled against the background of an influencing world, where the results of the dissidents' activities undulated through the structure holding the system together. The Fahrenheit Companions framework, once unassailable, started to give indications of weakness as additional people scrutinized its authenticity. The dissidents became images of a prospering transformation, motivating others to reexamine their own connections and challenge the imperatives forced by the Fahrenheit Companions scale.

As the development picked up speed, Alex and Mia ended up at the front of an unrest that looked to rethink the actual substance of human association. The renegades started to sort out get-togethers, both physical and virtual, where individuals could share their encounters and fashion kinships in view of certifiable seeing as opposed to cultural assumptions. The development rose above borders, interfacing people from different foundations who shared a typical craving for valid associations.

The Fahrenheit Companions Department, perceiving the danger presented by the developing development, heightened its endeavors to stifle disagree. The dissidents confronted expanding investigation and mistreatment as they kept on testing the laid out request.

The Agency conveyed trend setting innovation to screen and control relational connections, further fixing its grasp on the profound existences of the residents.

In spite of the heightening strain, the renegades proceeded, energized by the conviction that the innate warmth of human association couldn't be stifled by a framework intended to measure it. The development turned into an image of flexibility and trust, drawing in people from varying backgrounds who tried to break liberated from the shackles of fake companionship.

Amidst the developing opposition, Alex and Mia's own relationship extended. Their association filled in as a signal of realness in a world that undeniably esteemed similarity over certified feeling. Together, they became images of the upset, motivating others to scrutinize the limits put on their connections and to look for associations that went past the limitations of the Fahrenheit Companions scale.

As the dissidents got some decent momentum, breaks started to show up in the groundwork of the Fahrenheit Companions framework. Individuals began to scrutinize the erratic mathematical qualities relegated to their fellowships and understood the innate imperfections in a framework that tried to normalize something as mind boggling and nuanced as human association. The development started a social shift, empowering people to focus on better standards without compromise in their connections.

The renegades, presently an impressive power, conceived an arrangement to destroy the Fahrenheit Companions Department and free society from the requirements that had smothered veritable associations for a really long time. The upset finished in a progression of facilitated activities, both on the web and disconnected, as the renegades looked to destroy the mechanical foundation that had propagated the fake warmth of fellowships.

The fight between the revolutionaries and the Fahrenheit Companions Department arrived at its peak, with the two sides competing for command over the account of human association. The dissidents, furnished with the force of validness, confronted the Department's impressive arms stockpile of mechanical control and social molding. The destiny of a general public's personal scene remained in a critical state as the battle unfurled.

Amidst the confusion, Alex and Mia ended up defying the heads of the Fahrenheit Companions Department. The experience turned into a representative conflict between the powers of congruity and the dissidents who supported the untamed soul of veritable association. The heads of the Agency, once taken cover behind the façade of mathematical accuracy, presently stood uncovered, their command over the close to home existences of the residents disentangling.

As the revolutionaries made progress, the heads of the Fahrenheit Companions Agency endeavored to rescue their power by offering concessions and commitments of change. In any case, the renegades, having tasted the pleasantness of true association, were reluctant to think twice about. They requested a total destroying of the Fahrenheit Companions framework, pushing for a re-visitation of the natural idea of human connections.

The world looked as the defiance unfurled, and the shared perspective started to move. Individuals, motivated by the boldness of the revolutionaries, began to scrutinize the cultural standards that had bound their profound lives. The development turned into an impetus for change, igniting discussions about the genuine embodiment of kinship and the need to break liberated from the fake requirements that had characterized connections for such a long time.

In a climactic second, the renegades accomplished their objective. The Fahrenheit Companions Department was destroyed, and the mathematical scale that had administered kinships was annulled. The world, freed from the shackles of counterfeit warmth, embraced the unconventionality and wealth of true associations. The renegades, when untouchables, turned into the draftsmen of another time, where connections were characterized by veritable comprehension and shared encounters as opposed to mathematical qualities.

As the residue settled, Alex and Mia thought about the excursion that had driven them to this second. They understood that the genuine pith of kinship couldn't be estimated or bound to the furthest reaches of a Fahrenheit scale. It was a complicated interaction of feelings, shared encounters, and common grasping that resisted evaluation. The radicals, having won over the powers of similarity, remained as gatekeepers of this freshly discovered opportunity, guaranteeing that people in the future could never be limited by the requirements of fake kinship.

The world, presently freed from the harsh grasp of the Fahrenheit Companions framework, went through a significant change. Individuals embraced the magnificence of authentic associations, esteeming the profundity of connections over shallow mathematical qualities. The scars of the defiance filled in as a sign of the significance of realness in human collaborations, and society thrived in the recently discovered opportunity to frame associations in view of shared values, interests, and feelings.

In the result of the upset, Alex and Mia, alongside their kindred dissidents, became trailblazers of a development that reshaped the actual structure holding the system together. The idea of Fahrenheit Companions turned into ancient history, supplanted by an aggregate comprehension that real associations couldn't be normalized or bound as far as possible. The revolutionaries, once named as Anomalies, were presently celebrated as visionaries who had rocked the boat and freed society from the fake limitations of fellowship.

4.1 Introduce a diverse group of characters with temperature-related abilities.

In a world dissimilar to some other, where the texture of the truth was entwined with the components, a different gathering of people arose, each having novel temperature-related capacities. These exceptional people, known as Thermomancers, bridled the force of temperature control to shape their fates and explore the many-sided balance among intensity and cold.

At the core of this uncommon outfit was Ash, a searing and decided Thermomancer with the capacity to produce extreme intensity voluntarily. Ash's simple presence could warm the coldest of environmental factors, and flares moved readily available with an elegance that discussed both power and control. However, underneath the red hot outside, Ash held onto a profound feeling of obligation, utilizing their capacities to carry warmth and light to the haziest corners of the world.

Inverse Ash in the Thermomancer bunch was Frostbite, an emotionless and baffling person with the ability to order the chill of winter. Frostbite's touch could freeze water in a moment, and an air of cold encompassed them like an ethereal shroud. In spite of the frigid disposition, Frostbite conveyed a peaceful insight, figuring out the sensitive balance among intensity and cold that represented the Thermomancer domain.

An exuberant and wicked Thermomancer named Burst added a component of eccentricism to the gathering. Burst had the exceptional capacity to control both intensity and cold all the while, making a powerful interaction of components. Their jokes were frequently joined by eruptions of fire and unexpected chills, keeping the gathering honest and infusing a feeling of energy into their experiences.

In the midst of the Thermomancers, Breeze stood apart as the breeze whisperer, having the uncommon capacity to control air temperature. Breeze could call delicate breezes or savage blasts with a simple idea, and their presence conveyed an unpretentious warmth or reviving coolness relying upon their temperament. Breeze's association with the always moving breezes made them a fundamental part of the Thermomancer group.

Balancing the gathering was Gem, a Thermomancer whose capacities based on the glasslike excellence of ice. Gem could shape complex ice forms and make fragile ice designs with a simple touch. Their quiet disposition and propensity for making craftsmanship from frozen components brought a feeling of style and serenity to the Thermomancer aggregate.

All together, of Thermomancers left on an excursion that would test the restrictions of their capacities and manufacture tough bonds. Their singular assets and shortcomings made a unique collaboration that permitted them to defeat difficulties that no single Thermomancer could confront alone.

The world they possessed was a domain where the equilibrium of temperature held the way to opening secret secrets and untold powers. In their mission, the Thermomancers experienced old curios permeated with the quintessence of temperature, each introducing a novel test that necessary the aggregate abilities of the gathering to survive.

The primary relic they found was the Flameheart, a throbbing wellspring of extraordinary intensity concealed profound inside the core of a lethargic spring of gushing lava. Coal, attracted to the red hot energy, started to lead the pack in bridling the force of the Flameheart. As the gathering dug into the liquid caves, they confronted difficulties that tried their capacities as well as the bonds that kept them intact.

Frostbite, generally held and scrutinizing, tracked down a surprising strength despite searing temperatures. Burst, ever the epitome of turmoil, found a newly discovered center as they directed the blazing quintessence of the Flameheart. Breeze's breezes effectively cooled the environmental factors, keeping the gathering from surrendering to the abusive intensity, while Precious stone's ice models gave competitive edges in exploring the tricky territory.

Having effectively saddled the force of the Flameheart, the Thermomancers rose up out of the volcanic profundities more grounded and more receptive to the complexities of temperature control. Their excursion, in any case, was a long way from being done, as the curios they looked for were dispersed across different scenes, each introducing its own arrangement of difficulties.

The following curio, the Frostcore, looked for them in a domain ceaselessly covered in snow and ice. Frostbite, right at home, assumed responsibility for directing the gathering through the frozen scenes. The gnawing cold represented an alternate arrangement of difficulties, requiring the Thermomancers to adjust and use their capacities in imaginative ways.

Ash, commonly connected with heat, found the dormant potential in controlling temperature differentials. By controlling the differentiation between their red hot air and the freezing environmental elements, Ash made strong shockwaves that made ways through blizzards. Blast, delighting in the frigid territory, summoned stunning showcases of cold firecrackers that entranced and confused any expected enemies.

Breeze, sensitive to the unobtrusive vacillations in air temperature, recognized secret risks in the frozen scene, giving the gathering important premonition.

Gem's fondness for ice permitted them to make mind boggling structures that filled in as both cautious obstructions and vital vantage focuses. The Frostcore, when an impressive test, surrendered to the aggregate ability of the Thermomancers.

As the gathering advanced in their journey, they experienced a progression of relics that pushed the limits of their capacities. The Emberforge, an enormous chamber loaded up with steadily moving temperatures, expected the Thermomancers to synchronize their powers as a beautiful, unified whole. Ash's blazes moved pair with Frostbite's chilly rings, making a hypnotizing show of temperature dominance.

Burst, consistently unusual, tackled the fluctuating temperatures to make a stunning light show that filled in as both interruption and motivation. Breeze's breezes circled the surrounding temperature, forestalling outrageous spikes or drops that could undermine the fragile equilibrium inside the Emberforge. Gem's glasslike manifestations reflected and refracted the bunch tints of temperature, adding a hint of ethereal excellence to the procedures.

The Thermomancers, their aggregate capacities raised higher than ever, rose up out of the Emberforge with a more profound comprehension of the perplexing dance among intensity and cold. Their process turned into a demonstration of the strength of solidarity and the boundless conceivable outcomes that emerged when people with different gifts cooperated toward a shared objective.

However, the Thermomancers confronted their most considerable test when they found the subtle Balance Crown, a curio said to hold the ability to control the actual texture of temperature itself. The Crown was reputed to be concealed in an aspect where limits of intensity and cold coincided in a fragile harmony, and just the people who could explore this perplexing domain could expect to guarantee its power.

The excursion into the Harmony Aspect pushed the Thermomancers as far as possible. Ash's flares glinted and faded in the freezing cold, while Frostbite's chilly touch battled to grab hold in the singing intensity. Burst found their double control tried more than ever, as the harmony aspect requested a degree of equilibrium that rose above their standard tumultuous inclinations.

Breeze, the breeze whisperer, confronted unusual air flows that took steps to divert them into the limits. Gem's ice models dissolved and refroze in quick progression, requiring an exceptional degree of versatility. The Harmony Aspect turned into a pot that tried their capacities as well as the strength of their kinships.

Despite apparently outlandish difficulties, the Thermomancers found that the genuine force of the Balance Crown lay not in that frame of mind of temperature but rather together as one that could be accomplished between limits.

Ash and Frostbite, once saw as contrary energies, settled on something worth agreeing on in the comprehension that intensity and cold were two of a kind.

Blast, known for their eccentricism, figured out how to channel their tumultuous propensities into a controlled dance that embraced both intensity and cold. Breeze, receptive to the consistently moving breezes, found a newly discovered strength in the fragile equilibrium of temperature inside the Balance Aspect. Gem's dominance over ice turned into an image of versatility, as they etched unpredictable structures that caught the embodiment of balance.

As the Thermomancers effectively explored the difficulties of the Balance Aspect, they arrived at the core of the domain where the Harmony Crown anticipated. The relic, an indication of wonderful equilibrium, reverberated with the aggregate energy of the gathering. At the point when Ash, Frostbite, Blast, Breeze, and Gem contacted the Crown at the same time, a flood of force encompassed them, combining their capacities into an amicable ensemble of temperature control.

The Harmony Crown gave to the Thermomancers a degree of dominance that rose above individual restrictions. Together, they turned into the watchmen of harmony, guaranteeing that the fragile harmony among intensity and cold was kept up with in their reality. The relics they had accumulated, when images of challenge and win, presently filled in as courses for the Thermomancers' aggregate power.

Directly following their excursion, the Thermomancers, presently limited by an association that outperformed the simple control of temperature, kept on investigating the secrets of their reality. They became representatives of equilibrium, utilizing their capacities to carry concordance to districts where temperature limits took steps to upset the normal request. The different gathering of Thermomancers, when divergent people with extraordinary gifts, had produced a bond that rose above the restrictions of their natural powers.

The world, always different by the Thermomancers' mission, embraced the standards of equilibrium and solidarity. The temperature-related capacities that once separate people presently filled in as a demonstration of the excellence that could arise when various gifts were tackled as one. The Thermomancers, when a gathering of exceptional people, turned into an image of motivation for the individuals who looked to explore the intricacies of existence with beauty, understanding, and an agreeable harmony between the restricting powers that molded their reality.

4.2 The protagonist forms a team to tackle larger issues and mysteries in the world.

In a world overflowing with intricacies and secrets, the hero, a visionary and driven individual named Alex, wound up constrained to handle bigger issues that rose above individual capacities. Alex, outfitted with a sharp keenness and an unquenchable interest, perceived that a few difficulties required a cooperative methodology. In this way, an aggressive undertaking started — the development of a different and gifted group, every part offering exceptional abilities and viewpoints that might be of some value.

The main enroll to Alex's group was Maya, a splendid researcher with an unmatched mastery in state of the art innovation. Maya's logical psyche and capability in disentangling complex frameworks made her an important resource. Her creations and advancements could unwind the most unpredictable secrets, giving the group the innovative edge expected to explore the mysteries they were set to face.

Close to join the positions was Aiden, a magnetic and clever person with a foundation in strategy and exchange. Aiden's capacity to explore the unpredictable snare of legislative issues and assemble partnerships demonstrated critical in tending to worldwide difficulties. Their appeal and prudent methodology procured them the trust of assorted networks, preparing for cooperative endeavors that rose above borders.

Chasing a balanced group, Alex searched out Gabriel, a swashbuckler and pilgrim with a profound comprehension of old civilizations and failed to remember chronicles. Gabriel's information on archeological locales and social curios added a verifiable aspect to the group's capacities. His mastery was important in addressing secrets as well as in uncovering stowed away insights that could reshape the story of the world.

The group extended further with the expansion of Elena, a gifted language specialist and multilingual. Elena's capacity to translate old dialects and decipher obscure messages turned into an essential device in disentangling the mysteries that had escaped others for a really long time. Her semantic ability crossed over correspondence holes and opened ways to information that had extended stayed blocked off.

Finishing the gathering was Kai, a carefully prepared specialist and strategist with a foundation in military knowledge. Kai's capacity to evaluate circumstances, concoct exact game plans, and lead the group through complex provokes carried an upper hand to the gathering. Their authority abilities, sharpened in the pot of high-stakes situations, supplemented the different gifts of the group.

Together, Alex and this diverse gathering left determined to resolve bigger issues that tormented the world. The difficulties they confronted went from worldwide emergencies to antiquated secrets that resisted regular comprehension. The group, every part an expert by their own doing, consolidated their abilities to shape a firm unit that could handle complex issues head-on.

One of the group's earliest missions included researching a progression of unexplained peculiarities happening all over the planet. Maya's innovative skill permitted the gathering to break down examples and inconsistencies, while Gabriel's verifiable information gave setting to apparently inconsequential occasions. Elena's semantic abilities revealed old texts that indicated an association between the peculiarities and long-failed to remember fantasies.

As Aiden attempted to fabricate coalitions and assemble insight from conciliatory channels, Kai fostered a well defined course of action to address the unfurling secret. The cooperative endeavors of the group uncovered a secret organization controlling worldwide occasions for obscure purposes. Together, they formulated a countersystem, destroying the clandestine tasks and reestablishing harmony to the impacted locales.

Encouraged by their prosperity, the group directed their concentration toward additional critical worldwide difficulties. Environmental change, international strains, and financial incongruities arose as central focuses for their undertakings. Maya's innovative arrangements focused on feasible practices, Aiden's discretionary drives for global collaboration, and Kai's essential intending to resolve fundamental issues became key parts of the group's methodology.

Elena's examination into authentic points of reference gave bits of knowledge into how social orders had explored comparable difficulties previously. Gabriel's investigation of antiquated developments revealed failed to remember advances and practices that could illuminate present-day arrangements. The group, working cooperatively, introduced a thorough and comprehensive methodology to address the interconnected snare of worldwide issues.

Nonetheless, the group's process was not without its portion of misfortunes. They experienced opposition from dug in interests, confronted unanticipated results of their activities, and wrestled with the moral ramifications of their choices. The heaviness of the world's concerns overwhelmed them, testing the versatility of their coalition and the strength of their convictions.

Amidst these difficulties, Alex arose as the key part that kept the group intact. Their capacity to motivate, adjust, and lead with sympathy turned into a directing power for the gathering. Alex's vision stretched out past prompt critical thinking; they tried to encourage a shared awareness that rose above individual contrasts and joined the group in a common obligation to positive change.

The group's endeavors earned respect on a worldwide scale, drawing the consideration of persuasive figures, associations, and even states. While some embraced the cooperative methodology and perceived the requirement for aggregate activity, others saw the group as a danger to existing power structures. The division between the people who saw the group as heros and the individuals who saw them with doubt increased the stakes of their central goal.

Chasing unwinding worldwide secrets, the group uncovered a covert association with evil expectations. This shadowy gathering, working in the background, tried to control world occasions for their benefit. Maya's mechanical ability uncovered a snare of falsehood and secretive tasks, while Aiden's conciliatory artfulness uncovered the association's impact in political circles.

Gabriel's verifiable information followed the beginnings of this undercover gathering to old mystery social orders that had employed power since forever ago. Elena's phonetic abilities interpreted coded messages that uncovered the association's secret plans, and Kai's essential intuition fostered an arrangement to destroy their tasks and kill their impact.

As the group dug further into the core of the trick, they experienced difficulties that tried the constraints of their capacities and the strength of their bonds. The secret association, mindful of the danger presented by the group, released an influx of countermeasures and diversionary strategies. The group, confronting misfortune on various fronts, needed to recalibrate their procedures and adjust to an always advancing scene of duplicity and interest.

Notwithstanding these difficulties, the group found that their most noteworthy strength lay in their variety. Every part brought a novel arrangement of abilities and viewpoints, making a powerful cooperative energy that demonstrated strong notwithstanding misfortune. Their capacity to team up, convey, and trust each other turned into the bedrock of their prosperity, permitting them to explore the complex maze of the scheme.

The peak of their central goal unfurled in an emotional showdown with the heads of the surreptitious association. Maya's mechanical ability conflicted with the association's high level simulated intelligence frameworks, making a virtual landmark where the destiny of worldwide data remained in a precarious situation. Aiden's conciliatory exchanges uncovered the association's secret associations, unwinding the multifaceted trap of impact they had projected.

Gabriel, drawing upon the authentic information on antiquated secret social orders, went up against the pioneers with the heaviness of their own set of experiences. Elena's etymological abilities decoded scrambled messages that revealed the association's actual expectations, while Kai's essential brightness formulated an arrangement to destroy the association from the inside.

As the group stood up to the forerunners in a high-stakes standoff, the world watched, pausing its breathing. The disclosures uncovered by the group uncovered the loathsome ruses of the furtive association, prompting a worldwide retribution. The pioneers were considered responsible for their activities, and the group's endeavors introduced another period of straightforwardness and responsibility on the worldwide stage.

The result of the group's victory denoted a defining moment on the planet's impression of cooperative critical thinking. The group's prosperity turned into an encouraging sign, moving others to embrace the force of variety and joint effort in tending to complex difficulties. States, associations, and networks started to take on a more comprehensive and helpful methodology, perceiving that no single substance could explore the intricacies of the cutting edge world alone.

Alex, having driven the group through the maze of worldwide secrets, arose as an image of visionary initiative. Their obligation to solidarity, sympathy, and aggregate activity set a trend for another period of critical thinking. The group, having endured storms and prevailed over misfortune, became envoys of cooperative development, impacting positive change on a worldwide scale.

In the repercussions of their central goal, the group kept on cooperating, tending to new difficulties that arose in the always advancing scene of the world. Their excursion, set apart by the highs of progress and the lows of misfortune, turned into a demonstration of the groundbreaking influence of coordinated effort and the capacity of a different gathering of people to handle bigger issues that rose above the extent of any one individual. The tradition of their undertakings reverberated in the shared awareness, motivating people in the future to embrace the standards of solidarity, compassion, and cooperative critical thinking.

4.3 The group undergoes initial conflicts and learns to harness their collective powers.

Collecting a group with different gifts and foundations was not without its difficulties. At first, the gathering confronted clashes established in their disparities — clashes that would eventually turn into the cauldron through which they produced a more grounded, more brought together element.

Maya, driven by an immovable confidence in innovation, conflicted with Gabriel, whose trust lay in the insight of old civic establishments. Their conflicts turned into a microcosm of the more extensive battle between the cutting edge and the verifiable inside the group. Maya's dependence on information and investigation crashed into Gabriel's natural comprehension of the past, making pressures that took steps to subvert the cohesiveness of the gathering.

Aiden and Kai, with their particular ways to deal with critical thinking, wound up in conflict. Aiden, the representative, leaned toward nuanced dealings and sensitive moves in the domain of governmental issues.

Conversely, Kai, the tactical tactician, inclined towards conclusive activities and vital orders. The conflict between Aiden's artfulness and Kai's practicality turned into a wellspring of grating, featuring the differentiating ways of thinking inside the group.

Elena, the language specialist, frequently wound up trapped in the crossfire of clashing viewpoints. Her job in translating antiquated messages and deciphering messages expected her to explore the many-sided transaction of the group's elements. On occasion, she filled in as a middle person, endeavoring to connect the holes among Maya and Gabriel, Aiden and Kai, and other different matches inside the group.

The struggles, while testing, were not inconceivable. Alex, the visionary who had united the group, perceived the potential for development inside these strains. Rather than trying to eradicate the distinctions, Alex urged the group to embrace them as wellsprings of solidarity. The contentions, instead of being an obstruction, became open doors for the gathering to gain from each other and bridle their aggregate powers.

The defining moment came when the group confronted an emergency that expected their consolidated endeavors. A worldwide situation developed — startling and uncommon — that requested a quick and brought together reaction. The gathering, compelled to save their unseen fits of turmoil, mobilized together to resolve the bigger main thing.

Maya's mechanical experiences, Gabriel's verifiable information, Aiden's conciliatory associations, Kai's essential keenness, and Elena's semantic abilities combined in an orchestra of cooperation. The group, when isolated, found the groundbreaking force of their consolidated assets. The emergency turned into the impetus for a significant acknowledgment: their disparities, a long way from being hindrances, were the keys to opening their aggregate potential.

In the fallout of their effective reaction to the worldwide emergency, the group participated in transparent discussions. They perceived the significance of embracing variety and recognized that their special points of view were fundamental to handling the many-sided difficulties they confronted. Maya and Gabriel settled on some mutual interest in the combination of innovation and old insight, making a union that improved their critical thinking draws near.

Aiden and Kai, having endured their underlying conflicts, found the strength that arose out of the combination of strategy and key astuteness. Their cooperative endeavors turned into a model for exploring complex circumstances, outlining that a nuanced approach could coincide with definitive activity. Elena's job as a go between developed into a place of impact, as she directed the group towards an amicable equilibrium of viewpoints.

With the contentions behind them, the group entered another period of cooperation, where they tackled their aggregate powers with newly discovered collaboration. Maya's mechanical developments consistently incorporated with Gabriel's verifiable bits of knowledge, making a strong scientific structure that could disentangle even the most obscure secrets. Aiden's conciliatory artfulness supplemented Kai's essential brightness, guaranteeing that the group's activities resounded on both a worldwide and strategic scale.

Elena's phonetic abilities turned into the string that wove together the assorted embroidered artwork of the group. Her capacity to convey actually and cultivate understanding connected the holes between various characters and ranges of abilities. The contentions that once taken steps to destroy the group presently became wellsprings of solidarity, as every part contributed a one of a kind point of view to the aggregate knowledge of the gathering.

The group's process went on as they handled progressively complex difficulties. From revealing old curios with baffling powers to tending to international emergencies that compromised worldwide strength, the gathering moved toward each undertaking with a brought together front. Their cooperative endeavors turned into a signal of motivation, provoking others to perceive the extraordinary potential that lay in tackling different gifts and points of view.

Perhaps of their most many-sided mission included interpreting an old prescience that held the way to deflecting a disastrous occasion. Maya's mechanical examinations gave a contemporary setting, Gabriel's verifiable comprehension disclosed the beginnings of the prescience, Aiden's conciliatory artfulness associated the prediction to introduce day worldwide undertakings, Kai's essential sharpness conceived an arrangement for its execution, and Elena's etymological abilities disentangled the obscure messages concealed inside the old message.

As the group dug into the profundities of the prediction, they experienced difficulties that tried the constraints of their aggregate powers. The old text contained layers of imagery and similitude that expected a complex comprehension. Clashes, however definitely less articulated than previously, still emerged as the group wrestled with deciphering the subtleties of the prescience.

However, this time, the struggles filled in as impetuses for more profound comprehension. The group participated in cooperative conversations, drawing upon the lavishness of their different viewpoints to open the deeper implications inside the prescience. The contentions that once taken steps to prevent their advancement currently energized their scholarly interests, igniting experiences and leap forwards that carried them nearer to reality.

Simultaneously, the group found that their aggregate powers stretched out past the domains of innovation, history, tact, procedure, and semantics. The immaterial components of trust, sympathy, and shared vision turned out to be similarly essential in their quest for disentangling secrets and tending to worldwide difficulties.

Alex, the visionary who had imagined a unified and cooperative group, saw their underlying vision work out as expected in the amicable collaboration that had created among the individuals.

The peak of their main goal unfurled as the group went up against the looming devastating occasion anticipated by the antiquated prediction. The cooperative insight they had developed turned into their most noteworthy weapon. Maya's mechanical arrangements, Gabriel's verifiable bits of knowledge, Aiden's political exchanges, Kai's essential preparation, and Elena's etymological understandings merged in a planned exertion that deflected the catastrophe.

The outcome of their central goal denoted a zenith in the group's excursion. They had tackled their aggregate abilities as well as developed into a brought together power that rose above individual qualities and shortcomings. The contentions that had once taken steps to crash their endeavors had become venturing stones, impelling them towards a degree of coordinated effort that outperformed their underlying assumptions.

In the fallout of their victory, the group kept on handling bigger issues and secrets, furnished with the information that their aggregate powers were an awe-inspiring phenomenon. Their process turned into a demonstration of the extraordinary force of cooperation, showing that different viewpoints, when embraced and outfit successfully, could prompt arrangements that rose above the restrictions of individual capacities.

As the group wandered into new difficulties, they did as such with a mutual perspective that clashes were not to be dreaded however embraced as any open doors for development. The elements of their coordinated effort had developed from an assortment of capable people with varying perspectives to an agreeable ensemble of abilities, each adding to an aggregate orchestra that resounded with the potential for positive change on a worldwide scale.

The tradition of their process reverberated in the shared mindset, motivating others to perceive the strength that lay in joint effort and the power that arose when different gifts met up in quest for a shared objective. The group, having gone through introductory contentions and figured out how to bridle their aggregate powers, remained as an encouraging sign and a demonstration of the extraordinary potential that could be opened through solidarity and cooperation.

The excursion of outfitting aggregate powers turned into a characterizing section in the group's story. Having conquered introductory contentions and figured out how to see the value in the variety inside their positions, the gathering set out on a progression of missions that pushed the limits of their cooperative capacities. It was an excursion set apart by difficulties, disclosures, and the consistent development of the group into an imposing power equipped for resolving the most mind boggling issues confronting the world.

Their next mission unfurled in the shadow of an approaching natural emergency — a combination of environmental change, biological irregularity, and human exercises undermining the sensitive equilibrium of the planet. Maya's mechanical ability turned into a focal point of support in the group's reaction. She concocted state of the art answers for screen and moderate natural debasement, utilizing satellites, sensors, and progressed examination to accumulate continuous information.

Gabriel's verifiable bits of knowledge, established in the comprehension of how old civic establishments had coincided with nature, offered a reciprocal point of view. His insight directed the group in perceiving examples and arrangements that repeated the reasonable acts of the past. The conflict between Maya's dependence on mechanical development and Gabriel's respect for old insight changed into a cooperative exertion that consolidated the most ideal scenario.

Aiden's strategic artfulness assumed a pivotal part in preparing worldwide collaboration. Environmental change, as the group perceived, was a worldwide test that expected a bound together reaction. Aiden's capacity to construct unions and work with discussions between countries became instrumental in creating arrangements that rose above borders. Kai's essential sharpness, sharpened in military preparation, spread out an exhaustive technique for carrying out ecological strategies and drives on a worldwide scale.

Elena's semantic abilities demonstrated imperative in conveying the direness of the natural emergency to assorted networks all over the planet. Her capacity to verbalize the complex logical discoveries in a language that reverberated with individuals from various societies and foundations overcame any barrier between logical talk and public mindfulness. The struggles that once stewed underneath the surface were presently directed into an aggregate work to address the up and coming danger to the planet.

The outcome of their ecological mission exhibited the groundbreaking force of bridling aggregate powers. The group had developed past the amount of their singular abilities, showing that the crossing point of innovation, history, tact, system, and semantic correspondence could yield answers for difficulties that appeared to be unrealistic. Their cooperative reaction to the natural emergency turned into a plan for resolving worldwide issues through solidarity and shared mastery.

Floated by their prosperity, the group put their focus on revealing antiquated ancient rarities instilled with magical powers. This mission, established in Gabriel's verifiable information, drove them to antiquated sanctuaries, stowed away burial chambers, and neglected ruins. Maya's mechanical developments helped with filtering and examining these destinations, while Elena's etymological abilities translated mysterious engravings that directed the group further into the secrets of the past.

Clashes, albeit less articulated, still surfaced as the group wrestled with the obscure powers of the antiquities. Maya's mechanical methodology conflicted with the magical translations leaned toward by Gabriel and, somewhat, Elena. The antiquated relics, having energies that rose above customary comprehension, tried the group's capacity to accommodate their assorted points of view and convictions.

However, the struggles became open doors for the group to dive into the core of the secrets they tried to unwind. Cooperative conversations, powered by regard for one another's perspectives, prompted a union of innovation and otherworldliness. Maya's logical methodology uncovered the logical standards basic the curios' powers, while Gabriel and Elena brought a more profound comprehension of the social and otherworldly importance implanted in the old relics.

Aiden's political artfulness demonstrated priceless as the group experienced neighborhood networks and native gatherings associated with the antiquities. Building trust and cultivating joint effort became key parts of their main goal. Kai's essential keenness guaranteed the group explored the intricacies of both the physical and otherworldly domains, forestalling potentially negative results as they looked to bridle the powers inactive inside the relics.

The climax of their endeavors came in the revelation of a curio known as the Nexus Stone — a point of convergence of mysterious energies that could reshape reality. The contentions that had at first emerged according to contrasting viewpoints on science and otherworldliness were currently stations through which the group took advantage of the Nexus Stone's true capacity. Maya's mechanical connection point gave a way to comprehend and control the relic's energies, while Gabriel and Elena's experiences directed the group in saddling its powers dependably.

Aiden's discretionary associations demonstrated pivotal in drawing in with networks that held old information about the Nexus Stone, cultivating a cooperative exertion that crossed ages. Kai's essential arranging guaranteed that the group moved toward the antiquity with alert, understanding the fragile equilibrium expected to mindfully employ such magical powers. The Nexus Stone, when a wellspring of contention, turned into an image of the group's capacity to orchestrate their different assets.

Their process went on as they dove into the secrets of reality — a journey to open the mysteries of transient irregularities and layered fractures. Maya's mechanical advancements permitted the group to test the texture of the real world, while Gabriel's verifiable experiences gave setting to antiquated accounts of time travel and substitute aspects. The struggles that had at first emerged according to contrasting points of view on science and history presently energized a common interest with the unexplored world.

Elena's semantic abilities assumed a pivotal part in translating old texts that alluded to entrances between universes, while Aiden's strategic artfulness opened ways to joint efforts with specialists in quantum material science and hypothetical astronomy. Kai's essential sharpness, consistently capable at exploring strange domains, spread out an arrangement to investigate fleeting irregularities securely and comprehend the ramifications of controlling the texture of reality.

The struggles that surfaced during their investigation of reality were scholarly as well as moral. The group wrestled with the outcomes of adjusting verifiable occasions, the potential for accidental interruptions, and the moral ramifications of employing powers that rose above human comprehension. These contentions became moral compasses, directing the group as they continued looking for information while keeping a profound regard for the holiness of the past and the delicacy representing things to come.

The defining moment came when the group found a fleeting peculiarity that took steps to unwind the actual texture of the real world. Maya's innovative connection point, adjusted with accuracy, permitted the group to look into the oddity without upsetting its sensitive equilibrium. Gabriel's authentic bits of knowledge enlightened the likely results of intruding with the timetable, establishing the group in a consciousness of the dangers implied.

Elena's semantic abilities translated secretive admonitions implanted in antiquated texts, advised against foolish control of time. Aiden's strategic artfulness worked with discussions with worldly specialists and ethicists, making an aggregate system for moving toward the irregularity mindfully. Kai's essential intuition conceived an arrangement to settle the fleeting irregularity without modifying the direction of history.

Their progress in exploring the fleeting oddity turned into a demonstration of the group's development. The struggles that had once emerged from contrasting perspectives were presently essential to their capacity to move toward complex difficulties with subtlety, foreknowledge, and moral contemplations. The group had not just outfit their aggregate abilities in the domain of reality yet had additionally advanced into gatekeepers of the sensitive harmony among information and obligation.

The group's excursion, set apart by the nonstop development of their cooperative elements, turned into an account of flexibility, versatility, and the extraordinary force of solidarity. The contentions that had at first taken steps to wreck their endeavors had become venturing stones, pushing them toward a degree of joint effort that outperformed their underlying assumptions. As they faced bigger issues and secrets on the planet, the group exhibited that their aggregate powers were the amount of individual abilities as well as a power equipped for reshaping reality itself.

Directly following their achievements, the group turned into a motivation for others trying to cooperatively address complex difficulties. Their story reverberated as an encouraging sign, showing that the different crossing point of innovation, history, tact, technique, etymology, and morals could prompt arrangements that rose above the restrictions of individual capacities. The group's heritage reverberated in the shared mindset, a demonstration of the groundbreaking potential that could be opened through solidarity and joint effort.

Chapter 5

Celsius Conspiracy

It was 2057, and the world was wavering near the precarious edge of a devastating environment emergency. The impacts of many years of uncontrolled industrialization and natural disregard had at long last found humankind, and the planet was reeling under the heaviness of increasing temperatures, outrageous climate occasions, and waning assets.

In the midst of the bedlam, a gathering of researchers coincidentally found a stunning disclosure that would steer history. They revealed proof of a stealthy association known as the Celsius Consortium, a strong gathering with a vile plan. The Consortium had been controlling worldwide temperatures for quite a long time, coordinating a great intrigue to apply command over countries and store up incredible riches.

The researchers, drove by Dr. Olivia, still up in the air to uncover reality and ruin the Consortium's detestable plans. As they dove further into the scheme, they found a trap of misdirection and defilement that stretched out into the most noteworthy echelons of government, business, and the scholarly community.

The Celsius Consortium's usual methodology was both intricate and treacherous. They had created trend setting innovations fit for controlling atmospheric conditions, modifying sea flows, and in any event, impacting the World's turn. These controls permitted them to establish fake environment emergencies, which they took advantage of to unite power and control popular assessment.

Dr. Turner and her group confronted various difficulties as they attempted to beat the clock to unwind the complexities of the connivance. The Consortium was capable at covering its tracks, eradicating advanced impressions, and hushing question. The researchers needed to explore a slippery scene of disinformation, twofold specialists, and still up in the air to safeguard the mysterious that could shake the groundworks of the world request.

The way to uncovering the Celsius Connivance lay in translating a progression of obscure messages left by a previous individual from the Consortium who had turned informant. These messages, inconspicuous inside logical examination papers, monetary reports, and climate information, gave the breadcrumbs that would lead the group to the core of the intrigue.

As Dr. Turner and her partners dug into the secret universe of the Celsius Consortium, they uncovered an organization of shadowy figures with associations with large companies, compelling legislators, and worldwide associations. The Consortium's definitive objective was to control worldwide occasions, make fake emergencies, and benefit from the subsequent tumult.

The researchers before long understood that the very organizations intended to address environmental change and safeguard the climate were undermined by the Consortium's impact. Global environment arrangements, ecological guidelines, and manageable improvement drives were all under the shadow of the Celsius Trick. The acknowledgment sent shockwaves through mainstream researchers and lighted a need to get moving to uncover reality.

The more Dr. Turner and her group revealed, the more they became focuses of the Consortium's endeavors to smother their discoveries. They confronted arranged slanderous attacks, cyberattacks on their exploration, and, surprisingly, actual dangers. The informants who had at first connected with them were vanishing individually, abandoning a path of secret and dread.

In a test of skill and endurance, Dr. Turner and her associates jumbled the globe, sorting out the riddle of the Celsius Scheme. From secret gatherings in meeting rooms to undercover tasks in remote exploration offices, they followed the path of breadcrumbs, avoiding both clear and clandestine endeavors to upset their advancement.

The researchers likewise confronted unseen fits of turmoil inside their group as the extent of the intrigue became clear. Trust was hard to come by, and the apprehension about treachery waited in the air. In reality as we know it where loyalties were tried and partnerships were delicate, Dr. Turner needed to explore the outer dangers as well as the inside conflict taking steps to destroy their central goal.

As the proof mounted, the researchers gathered a dossier that uncovered the Celsius Trick in the entirety of its intricacy. The record point by point the Consortium's beginnings, its central participants, and its strategies for control. It additionally revealed the degree of the intrigue's effect on worldwide legislative issues, financial aspects, and natural strategies.

Furnished with the cursing proof, Dr. Turner and her group confronted the overwhelming assignment of exposing reality. They knew that uncovering the Celsius Trick wouldn't just break public trust yet in addition prompt a worldwide retribution with the truth of a controlled environment emergency. A lot was on the line, and the Consortium would remain determined to safeguard its insider facts.

The researchers formulated a painstakingly coordinated plan to deliver the data all the while through numerous channels, guaranteeing that it couldn't be smothered or excused as a simple fear inspired notion. They contacted confided in columnists, worldwide associations, and informant security gatherings, making a snare of collusions to enhance their message.

As the delivery date drew closer, strain mounted inside established researchers and among those in the loop. Murmurs of the looming disclosure flowed, making a feeling of expectation and tension. Legislatures and enterprises connected to the Consortium mixed to contain the aftermath, while customary residents wrestled with the ramifications of a world tricked by people with great influence.

The moment of retribution showed up, and the researchers opened up to the world about the proof of the Celsius Connivance. The disclosures sent shockwaves across the globe, setting off a flood of fights, requests for responsibility, and calls for fundamental change. The world had to go up against the truth that the very powers intended to address environmental change were complicit in its control.

States fell, corporate realms disintegrated, and the worldwide local area confronted a snapshot of retribution. The Celsius Trick uncovered the profundity of duplicity as well as the delicacy of the frameworks that represented the world. The disclosures provoked a reexamination of force elements, responsibility systems, and the job of science in forming strategy and public talk.

In the consequence of the confession, Dr. Turner and her group confronted both approval and judgment. A few hailed them as legends who had uncovered reality and ignited an upheaval, while others criticized them as disruptors who had weakened the fragile equilibrium of worldwide administration. The researchers became images of another time, one in which the quest for truth conflicted with the settled in interests of the strong.

As the world wrestled with the outcome of the Celsius Intrigue, questions emerged about how to remake trust, establishment straightforwardness, and manufacture a way toward certified ecological stewardship. The disclosures provoked a change in outlook in the manner in which society moved toward environmental change, driving a reevaluation of strategies, needs, and the actual idea of force.

In the years that followed, another time of worldwide participation arose. Countries met up to address the genuine difficulties presented by environmental change, unburdened by the controls of the past. The Celsius Scheme turned into a wake up call, an indication of the risks of unrestrained power and the significance of maintaining the standards of truth, responsibility, and ecological obligation.

Dr. Olivia Turner and her group, in spite of the individual forfeits and difficulties they confronted, became guides of motivation for another age of researchers, activists, and residents. Their process highlighted the force of truth despite trickery, the versatility of those able to rock the boat, and the aggregate strength of humankind when confronted with a typical danger.

The Celsius Scheme, when a shadowy phantom approaching over the world, was destroyed by the boldness and assurance of the individuals who wouldn't be controlled. As the world pushed ahead into an unsure future, the examples gained from the connivance filled in as an unmistakable update that the battle for truth and equity was a continuous undertaking, one that expected cautiousness, strength, and a promise to the overall benefit of mankind and the planet.

5.1 Uncover a sinister plot that involves manipulating temperature for personal gain or control.

Not long from now, the world ended up captured in the grip of a vile plot, a secret scheme that ventured into the actual texture of worldwide presence. The orchestrators of this pernicious plan were a shadowy association known as the Thermogenic Secrecy, a gathering of people driven by voracious eagerness, hunger for power, and an unfeeling dismissal for the prosperity of the planet and its occupants.

The Thermogenic Secrecy's plot based on an insidious arrangement to control temperatures on a worldwide scale. Their refined organization of researchers, technologists, and power merchants had conceived a strategy to apply command over the World's environment, making fake weather conditions and outrageous temperature variances to serve their tricky plan.

At the core of this connivance was a progressive innovation equipped for controlling climatic circumstances, sea flows, and, surprisingly, the World's center temperature. This innovation, created stealthily labs stowed away from intrusive eyes, empowered the Secrecy to impact the environment freely. The ramifications were faltering - dry seasons in prolific locales, destroying floods in bone-dry scenes, and the capacity to release mayhem with the flick of a switch.

The Secrecy's inspirations were twofold: individual increase and control. By controlling temperature and establishing counterfeit environment emergencies, they looked to design situations that would prompt asset deficiencies, financial breakdown, and boundless frenzy. In the following tumult, they expected to situate themselves as friends in need, offering arrangements and overseeing countries, state run administrations, and enterprises.

As the temperature control innovation became functional, the world started to encounter progressively serious and illogical climate occasions. Dry spells desolated once-ripe farmlands, while typhoons of remarkable power unleashed destruction on seaside urban areas. The Thermogenic Secrecy worked from the shadows, controlling the account through controlled news sources and impacting key figures in legislative issues and industry.

In the midst of this bedlam, a gathering of far-fetched legends arose. Dr. Elizabeth Hayes, a climatologist with a sharp feeling of instinct, started seeing themes in the environment information that didn't line up with normal cycles. As she dove further into her examination, she uncovered irregularities that pointed towards deliberate control instead of arbitrary events.

Dr. Hayes, driven by a feeling of obligation to mankind and the climate, gathered a group of specialists from different fields - meteorologists, PC researchers, and insightful columnists. Together, they left on a dangerous excursion to unwind reality behind the baffling environment moves and uncover the Thermogenic Secrecy's vile plot.

Their examination took them to the most obscure corners of the world, from cutting edge research offices concealed in distant mountains to meeting rooms where strong figures contrived to keep up with the state of affairs. The Secrecy, mindful of the danger presented by those trying to uncover their plot, conveyed an organization of specialists and moles to foil any efforts to uncover reality.

As Dr. Hayes and her group dug further into the trick, they experienced deterrents that tried their purpose. Cyberattacks designated their examination, eradicating significant information and undermining their correspondence channels.

The Secrecy's impact ventured into states, where key authorities chose not to see the mounting proof, influenced by the commitment of success and dependability presented by the shadowy association.

The legends of this unfurling show confronted individual penances also. Dangers to their wellbeing turned into a dependable friend, and trust inside the group faltered as the tremendousness of the trick became evident. The Secrecy, adroit at control and compulsion, took advantage of shortcomings inside the gathering, planting seeds of uncertainty and friction.

However, Dr. Hayes and her group went ahead, energized by an enduring obligation to uncover reality and shield the world from the hold of the Thermogenic Secrecy. They followed a path of breadcrumbs left by informants inside the Secrecy, people who had become frustrated with the association's mercilessness and were able to take a chance with all that to uncover reality.

The informants' disclosures gave basic experiences into the Secrecy's activities - the areas of their mysterious research facilities, the personalities of central members, and the degree of their impact over worldwide issues. Furnished with this data, the legends arranged to reveal the connivance to the world.

The excursion to uncover the Thermogenic Secrecy turned into a globe-running experience, with the group exploring a snare of misdirection that spread over mainlands. From the frosty scenes of the Icy, where the Secrecy maneuvered polar vortexes to plunge locales toward profound freezes, to the boiling deserts where misleadingly prompted heatwaves caused water sources to vanish, the legends followed the fingerprints of the Secrecy's obstruction.

As the proof mounted, the legends confronted a situation - how to uncover reality without setting off the very confusion the Secrecy tried to take advantage of. The response lay in essential collusions and a painstakingly organized plan to uncover the connivance all the while through different channels. They contacted confided in writers, natural associations, and thoughtful government authorities, making an organization of help to enhance their message.

The legends' endeavors finished in a worldwide disclosure that sent shockwaves all through society. The proof of the Thermogenic Secrecy's temperature control, carefully accumulated and introduced, uncovered the profundity of the trick. Legislatures fell, corporate realms disintegrated, and the world wrestled with the acknowledgment that it had been controlled for a terrific scope.

Public shock touched off fights across the globe, requesting equity and responsibility. The legends, when working in the shadows, presently ended up push into the spotlight as images of opposition against uncontrolled power. The Secrecy, its subtle pretense destroyed, confronted global judgment and a retribution for its violations against mankind and the planet.

In the repercussions of the disclosure, the legends confronted the overwhelming errand of reconstructing a world cracked by the Secrecy's ruses. Trust in foundations was broken, and the scars of the environment emergencies designed by the Secrecy ran profound. Legislatures, presently considered responsible by a stirred public, needed to diagram another course for natural approaches and worldwide participation.

Dr. Elizabeth Hayes and her group, when uncelebrated yet truly great individuals, became pioneers in the development to reestablish natural respectability and battle the impacts of the Secrecy's control. Their excursion from revealing an evil plot to starting a worldwide upheaval featured the force of truth, strength despite misfortune, and the limit of people to have an effect on the planet.

As countries cooperated to fix the harm fashioned by the Secrecy, another period of ecological stewardship arose. The innovation once utilized for evil objects was reused to improve the planet. Peaceful accords were fashioned to address environmental change all in all, with a freshly discovered obligation to straightforwardness, participation, and the protecting of the World's sensitive equilibrium.

The legends' heritage lived on as a demonstration of the victory of good over detestable, the versatility of the human soul, and the getting through force of truth. The world, scarred however stirred, pushed ahead into a future where the illustrations gained from the Thermogenic Secrecy's plot filled in as an obvious sign of the significance of cautiousness, responsibility, and the obligation we as a whole bear to safeguard the planet for a long time into the future.

5.2 The team embarks on a thrilling adventure to unravel the conspiracy.

In the faint shine of the examination lab, Dr. Olivia Rodriguez assembled her group of splendid personalities, each a specialist in their field. They were going to leave on an exhilarating experience that would test the constraints of their insight, boldness, and cooperation. The air was accused of expectation as Dr. Rodriguez divulged the proof they had carefully assembled - a path of breadcrumbs prompting a scheme that undermined their reality as well as the actual texture of the real world.

The intrigue, known as Undertaking Aegis, had stayed concealed for quite a long time, covered underneath layers of corporate mystery and government complicity. The group, a different gathering of researchers, programmers, and insightful columnists, had met up with a common mission: to reveal reality and uncover the powers controlling the world from the shadows.

The principal hint rose up out of an apparently harmless information break, a computerized break in the protective layer of a strong worldwide enterprise.

As the group dove into the taken records, they found a mind boggling trap of associations between compelling figures in business and governmental issues. It was a disclosure that indicated a vile plot with worldwide implications.

Dr. Rodriguez, a carefully prepared astrophysicist with a talent for interfacing divergent snippets of data, drove the charge. Her mastery in design acknowledgment and information examination turned into the key part of the group's analytical endeavors. As they filtered through monetary records, captured correspondences, and ordered reports, a more clear picture started to arise.

Project Aegis, it appeared, involved the control of key logical standards to adjust reality itself. The intrigue ventured into the most elevated echelons of force, with shadowy figures calling the shots in the background. The group's central goal became a scholarly pursuit as well as a test of skill and endurance to forestall the looming destructive occasion predicted in the taken documents.

Their process took them to the most distant corners of the globe, from clamoring cities to stowed away examination offices covered in remote scenes. Every area uncovered new layers of the intrigue, uncovering the degree of the control. The group confronted difficulties every step of the way - from corporate hired soldiers monitoring strictly confidential mysteries to not set in stone to smother reality.

As the group sorted out the riddle, they revealed the real essence of Undertaking Aegis. It wasn't simply an arrangement to control reality; it was a trial to bridle the actual texture of the universe. The schemers tried to use incomprehensible power, changing the laws of physical science to reshape the world as per their impulses.

The logical ramifications were faltering. Dr. Rodriguez, close by her splendid partners, deciphered complex conditions and obscure exploration papers into a story that even the layman could understand. The gravity of their discoveries turned into an energizing weep for the people who had been accidentally living in the shadow of Venture Aegis.

The group confronted outside dangers as well as inward difficulties also. Characters conflicted, self images conflicted, and trust turned into a valuable item. However, in the pot of their common mission, bonds were fashioned that rose above individual contrasts. Each colleague brought a remarkable arrangement of abilities and viewpoints, adding to the collaboration that impelled them forward.

The quest for truth turned into a high-stakes mental contest. The schemers, mindful of the group's endeavors, fought back with determined accuracy. Endeavors were made to ruin the group, eradicate their advanced impression, and even edge them for violations they didn't commit. The line among partner and enemy obscured as the group explored a misleading scene of double dealing and trick.

The defining moment came when the group revealed a secret research center settled in the core of a thick wilderness. It was here that the schemers directed their most brassy examinations, driving the limits of logical comprehension to the edge. As the group invaded the office, they confronted security frameworks intended to defeat even the most talented infiltrators.

Inside the research facility, the group found the climax of Venture Aegis - a gadget that tackled the force of quantum physical science to control reality itself. The backstabbers, driven by a craving for heavenly control, had planned to utilize this gadget to reshape the world as indicated by their vision. The ramifications were significant, and the group realized they needed to act quickly to forestall a disaster.

A strained a showdown resulted as the group went head to head against the schemers inside the tangled passageways of the secret office. It was a clash of brains and wills, with the destiny of the world yet to be determined. Dr. Rodriguez, outfitted with her logical keenness, participated in a verbal duel with the brains behind Undertaking Aegis, uncovering the defects in their rationale and the risks of their trial.

The climactic second shown up when the group figured out how to incapacitate the truth modifying gadget similarly as it arrived at a basic initiation point. The research facility, intended to contain the unfathomable powers at play, shuddered as the investigation's energy dispersed innocuously into the climate. The plotters, their stupendous arrangement obstructed, were captured and dealt with.

In the outcome of their triumph, the group confronted a world everlastingly different by the disclosures of Venture Aegis. Reality, uncovered so anyone might see for themselves, provoked a worldwide retribution with the harmony between logical advancement and moral obligation. State run administrations, organizations, and mainstream researchers had to think about the results of uncontrolled aspiration and the significance of moral limits.

Dr. Olivia Rodriguez and her group became legends, celebrated for their boldness despite existential dangers. The story of their experience enamored the public creative mind, motivating another age of researchers and scholars to move toward information with a feeling of obligation. Mainstream researchers, as well, went through a change in outlook, with expanded examination and straightforwardness turning into the standard.

As the world rose up out of the shadows of Venture Aegis, the group's process turned into an encouraging sign. Their cooperative soul, enduring assurance, and obligation to truth highlighted the force of people meeting up to defy difficulties that rose above the limits of science and society. The experience that started in the faint shine of an examination lab finished with a victory of keenness, mental fortitude, and the persevering through soul of humankind against the powers that looked to control the actual texture of the real world.

5.3 Face internal and external challenges that test the bonds within the group.

The group, having effectively obstructed the detestable Undertaking Aegis, wound up in the result of their reality changing triumph. In any case, as they relaxed in the help of having deflected a horrendous occasion, new difficulties arose, both inner and outside, that tried the very bonds that kept the gathering intact.

Inside, the elements inside the group became stressed. The serious idea of their central goal had produced profound associations, yet it had additionally uncovered separation points that took steps to break the solidarity they had developed. Characters conflicted, and the heaviness of the information they worried about made a concern that squeezed vigorously in each colleague.

Dr. Olivia Rodriguez, the astrophysicist who had driven the charge against Venture Aegis, ended up wrestling with the mental cost of the mission. The heaviness of obligation regarding uncovering and forestalling a reality-changing connivance had caused significant damage. Restless evenings and consistent tension had scratched lines all over, and the glimmer of triumph was tempered by the cost it had demanded on her prosperity.

Inside the group, relational struggles rose to the surface. The pressure of the mission had exacerbated strains that had been stewing underneath the surface. Contrasts in approach, clashing self images, and the result of dangerous circumstances prompted a tangible strain inside the gathering. The very characteristics that had made them a considerable group during the emergency presently took steps to destroy them.

As the group wrestled with inside difficulties, outside dangers lingered not too far off. The plotters behind Venture Aegis, however thwarted in their well conceived plan, were not ones to blur into haziness. The group turned into the objective of retaliations, with shadowy figures working in the outskirts to dishonor their accomplishments and stain their notorieties.

Tales circled in the media, stirring up misgivings about the realness of the group's discoveries. Murmurs of a fantastic deception or a scheme inside the intrigue flourished, planting seeds of uncertainty among the very individuals they had tried to safeguard. The outside challenges tried the group's flexibility as well as their capacity to explore a world that had a few lingering doubts of their thought processes and the truth they had uncovered.

Dr. Rodriguez, ever the pioneer, wound up in the unenviable place of overseeing both interior and outside pressures. The weight of initiative weighed vigorously on her shoulders as she looked to direct the group through the tempest of difficulties they currently confronted. The very characteristics that had made her a powerful pioneer during the emergency were presently scrutinized despite the group's developing elements and the outside powers neutralizing them.

The most important move towards tending to the inward difficulties included transparent correspondence. The group assembled to air their complaints, express their dissatisfactions, and go up against the issues that had been rotting. Dr. Rodriguez, perceiving the significance of keeping up with trust inside the gathering, cultivated a climate where each colleague felt appreciated and esteemed.

It became clear that the extraordinary idea of their central goal had made a bond produced in the cauldron of shared peril. Notwithstanding, a similar force had likewise amplified individual contrasts, prompting conflicts that required goal. The group, furnished with the acknowledgment that their solidarity lay in solidarity, focused on defeating their struggles under the surface and arising more grounded as a strong unit.

Remotely, the group confronted an advertising fight to rescue their standing. They drew in with confided in writers, researchers, and powerhouses to support their discoveries and counter the disinformation crusades pursued against them. The battle stretched out into the computerized domain, where the group's web-based presence turned into a milestone for truth against a scenery of deception.

The plotters, having pulled together in the shadows, sent a technique of unobtrusive damage. The group's exploration papers were investigated for any apparent irregularities, and their public appearances were taken apart for weaknesses. It was a round of insight, with the schemers utilizing their impact to stir up misgivings about the validness of the group's cases.

Accordingly, the group embraced straightforwardness. They opened their examination for autonomous investigation, permitting peer surveys and reviews of their systems. Dr. Rodriguez, standing firm in her obligation to truth, turned into the essence of the group's public guard. She participated in interviews, board conversations, and public discussions to address concerns and disperse the haze of suspicion that loomed over their disclosures.

The outer difficulties, in any case, negatively affected the group's psychological and close to home prosperity. The steady investigation, the flood of analysis, and the tenacious strain to demonstrate the veracity of their cases wore on them. However, with every difficulty, the group united behind a common perspective - the quest for truth and the guard of a reality that had been controlled for evil purposes.

In the midst of the bedlam, the group tracked down surprising partners. Researchers from different disciplines, columnists focused on analytical reporting, and, surprisingly, persuasive figures inside the tech business loaned their help. The story moved, and the tide started to change for the group as the heaviness of proof, peer supports, and public feeling moved towards acknowledgment of the truth they had uncovered.

At the same time, the group confronted secret dangers from leftovers of the Venture Aegis schemers. Endeavors were made to penetrate their computerized foundation, undermine their examination information, and even sow friction inside the gathering by taking advantage of their unseen struggles. The group, presently fight solidified and versatile, utilized online protection measures and counterintelligence strategies to foil these furtive endeavors.

Dr. Rodriguez, exploring the tempest on different fronts, tried to sustain the group's purpose. She underlined the significance of their central goal - as a logical pursuit as well as a reference point of truth in a world blurred by trickiness. The group's excursion, once restricted to the domain of examination labs and mystery offices, had turned into a worldwide adventure with suggestions that stretched out a long ways past the limits of their unique mission.

The defining moment came when an informant from inside the backstabbers' positions ventured forward. This courageous individual, driven by culpability and a longing for reclamation, gave indisputable proof of the backstabbers' endeavors to sabotage the group's validity. The disclosures uncovered the degree of the outer dangers and justified the group's cases.

Outfitted with this recently discovered proof, the group sent off a counteroffensive. They teamed up with insightful columnists to uncover the informant's story, making an account that illustrated the plotters' frantic endeavors to rescue their ruses. People in general, presently aware of the internal operations of the connivance, revitalized behind the group with reestablished energy.

The outside dangers started to scatter as the plotters, uncovered and cornered, withdrew into the shadows. The group's obligation to truth and their resolute assurance had beaten the powers that looked to control reality. The unseen struggles, as well, had provided way to a reestablished feeling of motivation and brotherhood manufactured in the cauldron of difficulty.

In the consequence of the tempest, the group arose as researchers as well as images of strength, respectability, and the victory of truth. Dr. Rodriguez, who had driven them through the most obscure minutes, remained at the front of an aggregate triumph against the powers that had looked to control the actual texture of the real world.

The world, having seen the group's excursion from uncovering a trick to beating inward and outside challenges, went through an aggregate change in discernment. The quest for truth turned into a common obligation, and the requirement for straightforwardness and responsibility notwithstanding control became vital.

Dr. Rodriguez and her group, once confronted with the overwhelming errand of protecting their discoveries, presently ended up celebrated as legends. Their excursion, set apart by inside struggle and outside dangers, turned into a demonstration of the force of strength, solidarity, and the immovable quest for truth notwithstanding difficulty.

As the group progressed from the power of their main goal to the quiet after the tempest, they thought about the illustrations learned. The bonds fashioned in the pot of their experience stayed whole, and the scars of their inner and outer fights became symbols of honor. The world, everlastingly different by the disclosures of Venture Aegis and the group's ensuing victory, pushed ahead into a future where the quest for truth remained as a core value against the shadows of control.

The consequence of their triumph against the treacherous Venture Aegis left the group in a condition of aggregate weariness, both genuinely and intellectually. The climate, once accused of the desperation of their central goal, presently balanced weighty with the heaviness of inner and outside challenges that tried the very bonds inside the gathering.

Inside the group, Dr. Olivia Rodriguez, the astrophysicist who had driven the charge against Undertaking Aegis, wound up wrestling with the repercussions of their serious mission. The cost of restless evenings, steady strain, and the obligation regarding uncovering and forestalling a reality-changing trick appeared in the lines carved all over. The triumph, however sweet, had included some significant pitfalls, and the cost it had taken on her prosperity was obvious.

The group, a different outfit of researchers, programmers, and insightful columnists, presently confronted the outcomes of their common difficulty. The inside elements, when a wellspring of solidarity, had become stressed. The pot of risk that had fashioned profound associations likewise uncovered separation points taking steps to break the solidarity they had developed during their central goal.

The first indications of inward difficulties arose in quite a while. The pressure of their high-stakes mission had exacerbated strains that had been stewing underneath the surface. Contrasts in approach, clashing self images, and the repercussions of dangerous circumstances prompted a discernible strain inside the gathering. The very characteristics that had made them an imposing group during the emergency currently took steps to destroy them.

Dr. Rodriguez, feeling the heaviness of initiative, perceived the direness of tending to the inner difficulties. A gathering was met, a chance for colleagues to air their complaints, express their dissatisfactions, and stand up to the issues that had been rotting underneath the surface. In the faintly lit room, enlightened exclusively by the sparkle of PC screens, each colleague talked, sharing their points of view and concerns.

It became obvious that the force of their central goal had made a remarkable bond - one fashioned in the cauldron of shared peril. Notwithstanding, a similar force had likewise amplified individual contrasts, prompting conflicts that required goal. The group, outfitted with the acknowledgment that their solidarity lay in solidarity, focused on conquering their struggles under the surface and arising more grounded as a durable unit.

Dr. Rodriguez, ever the pioneer, worked with transparent correspondence inside the gathering. She encouraged a climate where each colleague felt appreciated and esteemed. Trust, a delicate ware in the fallout of their central goal, turned into the focal point of their aggregate endeavors. The common objective of defeating struggles under the surface and saving the securities manufactured during their experience turned into a binding together power.

Remotely, notwithstanding, the difficulties kept on mounting. The backstabbers behind Task Aegis, however thwarted in their stupendous arrangement, were not ones to blur into lack of clarity. The group turned into the objective of backlashes, with shadowy figures working in the fringe to ruin their accomplishments and stain their notorieties.

Bits of hearsay circled in the media, causing some serious qualms about the realness of the group's discoveries. Murmurs of an excellent deception or a scheme inside the intrigue flourished, planting seeds of uncertainty among the very individuals they had tried to safeguard. The outside challenges tried the group's versatility as well as their capacity to explore a world that still had a few some lingering doubts of their intentions and the truth they had uncovered.

Dr. Rodriguez, presently exploring the tempest on various fronts, wound up in the unenviable place of overseeing both inward and outer tensions. The weight of initiative weighed intensely on her shoulders as she looked to direct the group through the difficulties they currently confronted. The very characteristics that had made her a viable pioneer during the emergency were currently scrutinized despite the group's advancing elements and the outside powers neutralizing them.

Remotely, the group confronted an advertising fight to rescue their standing. They drew in with confided in columnists, researchers, and powerhouses to confirm their discoveries and counter the disinformation crusades pursued against them. The battle stretched out into the computerized domain, where the group's web-based presence turned into a landmark for truth against a setting of falsehood.

The plotters, having refocused in the shadows, conveyed a procedure of unobtrusive damage. The group's exploration papers were examined for any apparent irregularities, and their public appearances were analyzed for weaknesses. It was a round of insight, with the schemers utilizing their impact to cause serious qualms about the credibility of the group's cases.

Accordingly, the group embraced straightforwardness. They opened their exploration for free investigation, permitting peer surveys and reviews of their philosophies. Dr. Rodriguez, standing firm in her obligation to truth, turned into the substance of the group's public guard. She participated in interviews, board conversations, and public gatherings to address concerns and scatter the haze of doubt that loomed over their disclosures.

The outside challenges, nonetheless, negatively affected the group's psychological and close to home prosperity. The consistent investigation, the blast of analysis, and the tenacious strain to demonstrate the veracity of their cases wore on them. However, with every mishap, the group came together for a mutual perspective - the quest for truth and the protection of a reality that had been controlled for terrible purposes.

In the midst of the tumult, the group tracked down surprising partners. Researchers from different disciplines, writers focused on analytical news-casting, and, surprisingly, powerful figures inside the tech business loaned their help. The account moved, and the tide started to change for the group as the heaviness of proof, peer supports, and public feeling moved towards acknowledgment of the truth they had revealed.

All the while, the group confronted incognito dangers from leftovers of the Venture Aegis plotters. Endeavors were made to invade their advanced foundation, undermine their exploration information, and even sow friction inside the gathering by taking advantage of their unseen fits of turmoil. The group, presently fight solidified and strong, utilized network protection measures and counterintelligence strategies to defeat these secret endeavors.

Dr. Rodriguez, exploring the tempest on numerous fronts, tried to invigorate the group's purpose. She stressed the significance of their central goal - as a logical pursuit as well as a reference point of truth in a world blurred by duplicity. The group's excursion, once restricted to the domain of exploration labs and mystery offices, had turned into a worldwide adventure with suggestions that stretched out a long ways past the limits of their unique mission.

The defining moment came when an informant from inside the plotters' positions ventured forward. This courageous individual, driven by culpability and a craving for reclamation, gave indisputable proof of the backstabbers' endeavors to sabotage the group's believability. The disclosures uncovered the degree of the outside dangers and justified the group's cases.

Equipped with this freshly discovered proof, the group sent off a counteroffensive. They teamed up with insightful columnists to expose the informant's story, making a story that illustrated the plotters' frantic endeavors to rescue their intrigues. General society, presently aware of the internal operations of the connivance, mobilized behind the group with recharged energy.

The outer dangers started to scatter as the schemers, uncovered and cornered, withdrew into the shadows. The group's obligation to truth and their unfaltering assurance had beaten the powers that looked to control reality. The unseen struggles, as well, had provided way to a recharged feeling of motivation and fellowship produced in the pot of misfortune.

In the repercussions of the tempest, the group arose as researchers as well as images of versatility, trustworthiness, and the victory of truth. Dr. Rodriguez, who had driven them through the haziest minutes, remained at the very front of an aggregate triumph against the powers that had looked to control the actual texture of the real world.

The world, having seen the group's excursion from uncovering a trick to defeating inner and outer difficulties, went through an aggregate change in discernment. The quest for truth turned into a common obligation, and the requirement for straightforwardness and responsibility notwithstanding control became principal.

Dr. Rodriguez and her group, once confronted with the overwhelming errand of guarding their discoveries, presently ended up celebrated as legends. Their excursion, set apart by interior struggle and outside dangers, turned into a demonstration of the force of strength, solidarity, and the unflinching quest for truth despite misfortune.

As the group progressed from the force of their main goal to the quiet after the tempest, they considered the illustrations learned. The bonds manufactured in the cauldron of their experience stayed solid, and the scars of their inward and outer fights became symbols of honor. The world, perpetually different by the disclosures of Task

Aegis and the group's ensuing victory, pushed ahead into a future where the quest for truth remained as a core value against the shadows of control.

Chapter 6

Kelvin's Redemption

In the modest community of Crestwood, where the nightfalls painted the sky in tones of orange and pink, resided Kelvin Hayes, a man tormented by the shadows of his past. Crestwood was where everybody knew every other person, where mysteries were intriguing, and the air held a specific serenity. In any case, Kelvin's life was not even close to tranquil.

His days were set apart by a tranquil urgency, a yearning for something he couldn't exactly eloquent. As the town's smithy, Kelvin was talented with his hands, producing cutting edges that sparkled like silver under the sun. However, his heart conveyed the heaviness of an untold story, one that murmured through the breezes of Crestwood.

Kelvin's excursion into the pit of despondency started quite a while back, a story woven with strings of disappointment and decisions that bore results heavier than the blacksmith's irons he worked with. His once good humored disposition had dulled, supplanted by an agonizing quiet that left the townsfolk pondering the man they assumed they knew.

Everything began with a commitment made on a blustery evening. The downpour flowed down in sheets, each drop repeating the strife inside Kelvin's spirit. That evening, he remained before the little church, looking for comfort from the whirlwind that reflected his conflict under the surface. The faint shine of the glimmering candles inside offered a similarity to warmth against the cool raindrops.

As Kelvin bowed in petition, a frantic request got away from his lips. The words were a blend of regret and misery, an admission to a more powerful that appeared to have neglected him. The commitment he made that evening would turn into the impetus for the series of occasions that would shape his predetermination.

The memory of her face tormented him — the lady he cherished, Elara. She was the girl of the town's minister, a reference point of honesty in a world that had become progressively pessimistic. Their affection was prohibited, a mystery murmured in obscurity corners of Crestwood. Be that as it may, energy, far and wide, consumes without leniency.

The surreptitious gatherings, taken looks, and the taken snapshots of joy turned into the material on which Kelvin painted his destruction. The townsfolk, speedy to pass judgment, cast disliking glares upon the couple. Elara's honest's dad, a harsh man, considered the association to be blasphemy. The strain escalated as murmurs transformed into allegations, and Kelvin wound up at a junction.

Confronted with the final proposal of leaving his affection or confronting expulsion from the main home he had at any point known, Kelvin settled on a decision that would resonate as the years progressed. In a snapshot of shortcoming, he picked self-protection over adoration. Elara, grief stricken and sold out, looked as Kelvin left, abandoning broke dreams and an affection that would not bite the dust.

The years that followed were a haze of repetitive daily schedule for Kelvin. He hurled himself entirely into his work, the musical crashing of his sledge against metal filling in as a mechanical cradlesong that muffled the reverberations of his past. Crestwood, while pardoning in nature, couldn't delete the scars scratched on Kelvin's spirit.

At some point, a letter showed up — a solicitation to the wedding of Elara. The news struck Kelvin out of nowhere, reigniting the ashes of his covered feelings. Angrily and franticness, he took to the container, looking for asylum at the lower part of each glass. The townsfolk, thoughtful however vulnerable, looked as Kelvin dropped into a purposeful limbo.

As the day of Elara's wedding drew nearer, Kelvin's inward storm arrived at its apex. A tempest blended both inside and outside, reflecting the disorder in his heart. Unfit to bear the heaviness of his choices any longer, Kelvin chose to stand up to his evil spirits and go to the wedding — a compensation he considered significant for the recovery he looked for.

The congregation chimes rang gravely as Kelvin moved toward where he had once made a commitment, and where he had broken another. The air was thick with pressure as he ventured into the blessed lobbies, his eyes meeting those of Elara, who stood brilliant in her wedding outfit. Time appeared to stop as their looks locked — a quiet trade of words implicit.

The service unfurled like a strange dream, Kelvin a simple observer to an affection he had once been a piece of. As Elara traded promises with another, Kelvin felt the heaviness of his transgressions overwhelm him. The walls of the congregation, observer to his quiet admissions, appeared to shut in.

In the consequence of the function, Kelvin remained external the congregation, the downpour washing away the leftovers of his previous self. The townsfolk, presently thoughtful to his predicament, watched from a good ways. Elara moved toward him, a combination of bitterness and figuring out in her eyes. She expressed words that conveyed the heaviness of remission, excusing him for the aggravation he had caused.

Kelvin, in any case, tracked down no comfort in her absolution. The weight of his decisions stayed, a weighty chain that bound him to a past he was unable to get away. Not entirely settled to break free, he set out on an excursion of self-disclosure, a journey for reclamation that would take him to the most obscure corners of his spirit.

His movements drove him through barren scenes and failed to remember towns, each stage a difficult sign of the distance he still couldn't seem to cover. En route, Kelvin experienced people whose accounts repeated his own — broken spirits looking for comfort even with their own slip-ups. The experiences filled in as a mirror, mirroring the cracked bits of his own character.

In one town, he met an isolated craftsman who painted the desolation of lost love on materials that appeared to drain with feeling. In another, a shrewd elderly person shared stories of pardoning and fresh opportunities, offering Kelvin a hint of something to look forward to in the most obscure of times. Each experience added a layer to how he might interpret reclamation, an idea that demonstrated subtle.

Tormented by the apparitions of his past, Kelvin confronted snapshots of shortcoming and allurement. The appeal of commonality, the solace of the jug, and the reverberations of his previous life took steps to crash his excursion. However, he went ahead, driven by an internal fire that would not be doused.

As Kelvin navigated the scenes of his own reclamation, he coincidentally found a religious community settled in the mountains. The priests, attendants of old insight, greeted him wholeheartedly. In their middle, Kelvin tracked down a coach — a shrewd priest whose eyes bore the heaviness of hundreds of years. Under his direction, Kelvin dug into the lessons of self-absolution and the craft of giving up.

The religious community turned into a safe-haven, where Kelvin stood up to his inward evil spirits with the direction of old ceremonies and contemplation. The interaction was burdensome, every disclosure a difficult sign of the injuries he had caused upon himself as well as other people. However, through the murkiness, he witnessed a gleam of light — a flicker of the reclamation he looked for.

Kelvin's excursion, both physical and otherworldly, turned into a demonstration of the human limit with regards to change. The scars of his past changed into identifications of strength, and the aggravation that once characterized him turned into the impetus for development. The townsfolk of Crestwood, far off observers to his change, started to detect an adjustment of the air — a change in the breezes that murmured stories of reclamation.

Getting back to Crestwood, Kelvin found the town unaltered yet unique. The air, when weighty with judgment, presently conveyed a breeze of acknowledgment. Individuals, when vigilant, broadened tokens of altruism. Kelvin, nonetheless, realize that genuine reclamation was an excursion without a characterized endpoint. It was a ceaseless pattern of development and expiation, a pledge to carry on with a day to day existence that respected the examples learned.

Elara, presently a widow with a kid, moved toward Kelvin with a reluctant grin. The injuries of the past, while not neglected, had mellowed with time. The two common a discussion that rose above words — a quiet affirmation of the common aggravation that bound them. At that time, Kelvin understood that reclamation was not tied in with eradicating the past but rather figuring out how to coincide with it.

The town of Crestwood, when an observer to Kelvin's fall out of favor, presently remained as a scenery to his continuous recovery. The smithy's fashion, when a position of singular work, turned into a center point of local area and shared stories. Kelvin, presently not characterized by the slip-ups of his childhood, arose as a mainstay of solidarity for those looking for their own way to reclamation.

The seasons changed, and Crestwood saw the rhythmic movement of life. Kelvin's excursion, while profoundly private, resounded with the townsfolk in manners he could never have expected. His story turned into an illustration — an update that recovery was not an objective but rather a persistent work to transcend one's past.

As the years unfurled, Kelvin's presence in Crestwood became inseparable from flexibility and trust. The scars on his spirit, when crude and difficult, changed into a wellspring of motivation for others confronting their own devils. The metalworker, once spooky by shadows, stood tall as a reference point of reclamation — a living demonstration of the extraordinary force of renewed opportunities.

In the calm snapshots of reflection, Kelvin understood that his reclamation was not a secluded excursion but rather a wave that contacted the existences of people around him. The embroidery of Crestwood, woven with strings of delight and distress, bore the signs of a man who would not be characterized by his mix-ups. The townsfolk, once doubtful, presently shifted focus over to Kelvin as a wellspring of motivation — an undeniable evidence that reclamation was feasible, even in the most obscure of times.

As Kelvin's story interlaced with the texture of Crestwood, the actual town went through an unpretentious change. The air, when weighty with judgment, presently reverberated with the tunes of pardoning. The congregation, where Kelvin had made and broken guarantees, remained as an image of elegance and acknowledgment. The downpour that had once washed away his tears presently supported the roses that sprouted in the town square — a similitude for the resurrection that recovery brought.

In the sundown of his years, Kelvin tracked down comfort in the straightforwardness of life. The once fierce soul currently embraced the serenity of the current second. The metal forger's sledge, once employed with a feeling of obligation, presently hit with a musical rhythm that repeated the concordance inside. Crestwood, an observer to Kelvin's excursion, remained as a living demonstration of the force of recovery to mend even the most profound injuries.

Thus, in the blurring light of day, Kelvin Hayes, the metal forger of Crestwood, thought back on his life — an embroidery woven with strings of despondency and recovery. His story, a section in the records of a modest community, filled in as an update that the excursion to recovery was a winding way, full of difficulties and wins.

As the sun plunged underneath the skyline, creating long shaded areas over the scene, Kelvin realize that his reclamation was not a determination but rather a continuation — a story that would be gone down through ages. The reverberations of his process would wait in the breezes of Crestwood, murmuring stories of an all man, despite everything, tracked down recovery in the tranquil corners of his heart.

Thus, the town of Crestwood rested under the vigilant look of the stars, the tradition of Kelvin's recovery carved into its actual establishment. The smithy's produce, presently an image of strength and change, remained as a quiet demonstration of the persevering through force of the human soul to miraculously rise like a phoenix after its own errors. In the quietness of the evening, as the world slept, the reverberations of Kelvin's recovery resonated through the embroidery of Crestwood — an immortal story of trust, pardoning, and the persevering through quest for another opportunity.

6.1 Dive into the backstory of a seemingly antagonistic character, revealing their motivations and struggles.

In the clamoring city of Eldoria, where transcending high rises penetrated the sky and the murmur of action resonated through the jam-packed roads, carried on with a man named Adrian Blackthorn. Known for his sharp keenness and heartless business discernment, Adrian was a figure covered in secret and frequently saw as a main bad guy in the perplexing embroidery of Eldoria's power elements.

Adrian's ascent to unmistakable quality was however quick as it might have been perplexing. From the shadows, he explored the perplexing snare of corporate governmental issues, leaving a path of partnerships and foes afterward. The media painted him as a savage magnate, a manikin ace making things happen in secret. In any case, behind the facade of the cool, computing money manager lay a story of strength and a past set apart by difficulties.

Adrian's process started in the edges of Eldoria, in a local where decrepit structures bore the scars of disregard. His young life was defaced by destitution, his family battling to earn enough to get by. Since early on, Adrian showed a sharp insight, a sharp brain that saw potential open doors where others saw just deterrents. It was this astuteness that became the two his approval and his revile.

As a young person, Adrian rioted, utilizing his brains to make due in the unforgiving metropolitan scene. He became snared with a nearby pack, a shoddy group of mavericks and untouchables who found comfort in the kinship brought about for a specific need. It was inside this world that Adrian leveled up the abilities that would later turn into his instruments for exploring the merciless corporate field.

Life in the group was a consistent battle for endurance, a determined hit the dance floor with risk and hardship. Adrian, in any case, saw it as a preparation ground — a cauldron that molded his personality and showed him the craft of transformation. His sharp brain turned into a weapon, a way to outfox opponents and gain a high ground in the unforgiving roads.

In those early stages, Adrian confronted treachery and misfortune. The roads guaranteed companions and partners, leaving scars that ran profound. The cruel truth of his environmental factors filled a consuming aspiration inside him — an enduring longing to get away from the grip of neediness and make an existence of force and impact.

Adrian's chance arrived in a startling structure. An opportunity experience with a well off business visionary made the way for a world he had just seen from a remote place. The business person, intrigued by Adrian's acumen and crafty, offered him an apprenticeship — a break from the shadows of the roads into the amazing lights of the corporate world.

Under the mentorship of his promoter, Adrian flourished. He ingested information like a wipe, learning the complexities of business, money, and procedure. The corporate world, with its ferocious rivalry and confounded governmental issues, ended up being not so not the same as the roads he had come from. Adrian, be that as it may, was as of now not an outcast. He was a chameleon, consistently mixing into the world he once tried to overcome.

As the years passed, Adrian ascended the company pecking order with a practically mysterious speed. His insight and mercilessness procured him a standing that struck trepidation into the hearts of his rivals. The media depicted him as a bad guy, a man who might persevere relentlessly to accomplish his objectives. Nonetheless, reality ran further than the titles proposed.

Underneath the cleaned outside of the industry mogul snuck the shadows of his past. Adrian's tireless quest for progress was not exclusively determined by voracity or a hunger for power; it was a mission for approval, a longing to demonstrate that he could rise above the conditions of his childhood. The scars of neediness, the misfortunes experienced on the unforgiving roads, energized a fire inside him that wouldn't be smothered.

Adrian's own life was an embroidery of penances. Connections disintegrated under the heaviness of his desires, kinships dissolved by the requests of his profession. The dejection that went with his rising turned into a quiet sidekick, a steady indication of the cost he paid for progress. The manor he lived in, the extravagant way of life he drove — every image of win was likewise a demonstration of the isolation he was unable to get away.

One vital second in Adrian's life cast a long shadow on his way. The demise of his more youthful sister, Lily, in a terrible episode shook the underpinnings of his apparently impervious outside. Lily had been the guide of light in Adrian's life, a wellspring of warmth in the chilly world he possessed. Her inauspicious downfall left a void that no measure of riches or influence could fill.

Distress turned into Adrian's quiet buddy, a ghost that spooky the edges of his psyche. The misfortune, as opposed to breaking him, energized a persistent drive to make a reality where others wouldn't need to confront the difficulties he and Lily had persevered. Underneath the facade of the merciless finance manager, Adrian held onto a longing to utilize his leverage for positive change — a mission powered by the memory of a sister he was unable to save.

Eldoria, for all its transcending high rises and shining veneers, held onto its own murkiness. Debasement, imbalance, and wrongdoing were the underside of the city's flourishing. Adrian, having seen the brutal real factors of life, tried to destroy the frameworks that propagated this murkiness. His generous undertakings, frequently disregarded by the media focused on his more dubious dealings, expected to elevate the neglected corners of Eldoria.

Adrian's double personality — a heartless financial specialist in the public eye and a quiet promoter in secret — was an unstable difficult exercise. The public saw him as a main bad guy, an image of corporate insatiability. Much to their dismay that similar man they attacked was discreetly attempting to change the very framework he was essential for.

The corporate scene of Eldoria was a front line where collusions were delicate, and double-crossings were normal. Adrian explored this world with an accuracy that ruled out feeling. The very insight that had saved him on the roads currently impelled him to pursue key choices that accumulated both profound respect and ill will. Eldoria, in any case, was a city that flourished with uncertainty, and Adrian Blackthorn was its baffling puppeteer.

Adrian's inspirations were perplexing, an embroidery woven with strings of desire, sadness, and a longing for approval. The city that he tried to change opposed change, its underlying foundations implanted in a past filled with epic showdowns and fundamental defects. Adrian, in any case, was an awe-inspiring phenomenon, a man who comprehended that genuine change required destroying the very establishments that maintained business as usual.

His charitable undertakings focused on schooling, medical care, and social change were met with opposition from the individuals who benefitted from the current differences. Adrian's endeavors to steer the results for the underestimated painted an objective on his back. The media, fast to benefit from debate, zeroed in on the conflicts as opposed to the fundamental mission.

Adrian's own battles, taken cover behind the emotionless exterior, arrived at a limit when he confronted a corporate trick pointed toward discoloring his standing. Allegations of misappropriation and defilement surfaced, taking steps to fix all that he had worked for. The media, consistently hungry for an outrage, enhanced the charges, giving Adrian a role as the encapsulation of corporate villainy.

The fight that followed was for Adrian's standing as well as for the goals he supported. The city looked as he battled like the devil to demonstrate his innocence, uncovering a trap of double dealing that arrived at the most noteworthy echelons of Eldoria's tip top. It was a fight that drove Adrian to the edge, driving him to stand up to the evil spirits he had long kept under control.

Amidst the unrest, Adrian's past reemerged unexpectedly. Individuals from his previous group, leftovers of the existence he had battled to abandon, moved toward him looking for help. Eldoria's underside, with its appendages arriving at even the most rich passages of influence, took steps to consume the individuals who couldn't get away from its grip. Adrian, conflicted between his over a significant time span, confronted a quandary that tried the restrictions of his convictions.

The disclosure that the trick against him was organized by strong figures inside Eldoria's corporate and political scene just filled Adrian's assurance. The fight for his standing changed into a campaign for equity — a mission to uncover the decay at the center of the city's power structures.

The hardships Adrian looked during this wild period uncovered weaknesses that the general population had only occasionally seen. The man once apparent as a savage big shot was, as a general rule, a person troubled by the heaviness of his decisions. The scars of his past, the deficiency of his sister, and the constant quest for a superior world merged in a powerful coincidence of internal conflict.

As reality behind the intrigue disentangled, the public's view of Adrian went through a progressive shift. The once-defamed figure turned into an image of versatility, a man who dealt with difficulty directly as he continued looking for equity. Eldoria, while still distrustful, started to perceive the duality of the one who had for quite some time been a secretive conundrum.

In the fallout of the uncover, as the plotters confronted the outcomes of their activities, Adrian remained at a junction. The city that had seen his climb and the ensuing fall currently anticipated his best course of action. The injuries of the fight were still new, however so was the chance for change.

Adrian, at this point not satisfied with simply exploring the framework, put his focus on destroying it. His charitable endeavors picked up new speed as he directed the public opinion towards an aggregate craving for a superior Eldoria. The media, when an enemy, turned into a partner in spreading the message of change.

The one who had been seen as a main bad guy arose as an encouraging sign — a demonstration of the extraordinary force of strength and the unflinching quest for a respectable goal. Adrian Blackthorn, when a name inseparable from corporate mercilessness, became inseparable from the chance of recovery and change.

The city of Eldoria, formed by the situation that unfurled inside its limits, saw a transformation. The shadows of defilement started to subside, and the seeds of a more fair society were planted. Adrian, presently a nonentity for change, kept on exploring the intricacies of his double presence, mindful that genuine change required in excess of a solitary man's endeavors.

As Eldoria embraced the beginning of another period, Adrian Blackthorn stayed a puzzling figure — a man whose past, present, and future were interwoven with the city he looked to change. The battles, penances, and recovery of one apparently hostile person turned into a demonstration of the multifaceted dance among power and profound quality, desire and charitableness, in the consistently developing embroidery of Eldoria.

6.2 Develop empathy for the antagonist, adding depth to the overall narrative.

In the core of a curious town settled between moving slopes and wandering streams, there carried on with a man named Victor Belmont. To the residents, he was a conundrum, a figure covered in secret and shadows. Victor was the main bad guy in the unfurling story of the town's battles, a man whose activities bore the heaviness of both hatred and interest.

His standing went before him, a story woven from the strings of dread and disdain. Victor, with his overwhelming height and penetrating look, turned into the exemplification of the town's aggregate bad dreams. Local people talked about dull dealings, condemnations, and mishaps that appeared to continue afterward. It was simple for the locals to give him a role as the main bad guy, an outcast whose presence undermined the congruity they held dear.

In any case, behind the exterior of the unfavorable figure was a story that unfurled like a sad embroidery. Victor's underlying foundations in the town ran profound, weaved with a background marked by familial love and double-crossing. As a young fellow, he had been the intemperate child of the Belmont family, regarded and respected for his knowledge and sympathy. Nonetheless, destiny had a horrible feeling of incongruity, and misfortune struck while an overwhelming sickness killed his folks.

Passed on to bear the weight of despondency and obligations, Victor's life veered off in a strange direction. The locals, when partners, betrayed the Belmont family in their period of scarcity. The murmurs of treachery, powered by dread and vulnerability, painted Victor as the harbinger of adversity. It was in this cauldron of misery that the seeds of hatred flourished.

With the heaviness of the world on his shoulders, Victor looked for comfort in the prohibited expressions — old customs and spells went down through ages. The locals, fast to pass judgment and ease back to grasp, considered these pursuits to be malignant, a plummet into obscurity. Unbeknownst to them, Victor's goals were not vindictive; they were conceived out of franticness, a longing to reverse the situation of setback that gripped to his family like a wrathful soul.

As Victor dove further into the illegal expressions, the town walked out on him altogether. The once-extravagant child turned into an outsider, an untouchable whose each step was met with doubt. The disconnection weighed vigorously on Victor, his franticness appearing in a fixation to recover the regard and nobility that had been taken from him.

The defining moment in Victor's life came as a prohibited book — a grimoire murmured to hold the way to unbelievable power. The murmurs of the town, the allegations of foul play, and the weight of his family's heartbreaking inheritance filled his assurance. Victor, with a heart weighty with misery and a psyche blurred by distress, capitulated to the charm of the dull expressions.

The ceremonies he performed were frantic endeavors to overcome any barrier between the human and otherworldly domains. His expectations, however brought into the world from a position of distress, became wound by the very powers he tried to control. The town, uninformed about the subtle conflicts that consumed Victor, saw just the outside indications of his drop into dimness.

As the town endured a progression of disasters — bombed crops, puzzling diseases, and unexplained hardships — the fault fell solidly on Victor's shoulders. The once-extravagant child turned into the substitute for each misfortune that happened to the local area. The locals, dazed by dread and hatred, neglected to see the man behind the apparent perniciousness.

Victor, troubled by the heaviness of the town's contempt, proceeded with his single interests. The grimoire, with its obscure mantras and eldritch images, became the two his salvation and his revile. Every custom played out, every frantic endeavor to modify the direction of destiny, drove him further into a pattern of misery and fixation.

The awfulness of Victor's story lay in the certainty of his separation. The more he looked to reclaim his family's name, the further he slipped into the pit of the mysterious. The residents, unwavering in their judgment, neglected to see the humankind that waited inside the tortured man. They couldn't grasp the interior fights that seethed inside him — the contention between a child's adoration and the charm of taboo power.

In a touch of destiny, a solitary vagabond showed up in the town — an explorer whose appearance bore the commitment of progress. This more peculiar, unburdened by the biases that tormented the locals, found in Victor a man tormented by his past and tormented by his decisions. The drifter, through sympathy and understanding, tried to unwind the intricacies of Victor's story.

As the drifter dove into Victor's past, the layers of the story started to strip away, uncovering a story of misfortune, disloyalty, and a frantic mission for reclamation. Victor, when seen as the exemplification of malice, turned into a heartbreaking figure — a man pushed to the edge by the savageries of destiny and the unforgiving judgment of his local area.

The town, through the eyes of the vagabond, turned into an impression of the more extensive human inclination to fear the unexplored world. The murmurs of condemnations and the allegations of dim expressions were systems of self-safeguarding, manners by which the residents limited any association with the awkward truth — that Victor's story could undoubtedly be their own assuming destiny were less lenient.

The vagabond, outfitted with sympathy and a longing for truth, looked to overcome any issues among Victor and the town.

Discussions were had, insights were uncovered, and the once-quiet bad guy turned into a man with dreams and desires that rose above the shallow marks forced upon him. The drifter, by disentangling the layers of Victor's story, provoked the residents to defy their own predispositions and biases.

In a climactic second, as the town wavered near the very edge of irreversible judgment, the drifter introduced reality — the story of Victor Belmont, a man whose activities, however misinformed, originated from a position of significant distress and the frantic craving to recover what had been lost. The grimoire, when seen as a device of malice, turned into an image of a man's battle against the unyielding powers of fate.

The disclosure ignited a change inside the town — an aggregate acknowledgment that the genuine bad guy was not Victor but rather the trepidation that had held them hostage. The vagabond's presence turned into an impetus for change, an update that sympathy and understanding could destroy the obstructions that isolated them from each other.

As the locals faced their own biases, they settled on some shared interest with Victor. The once-excluded man, through shared encounters and a recently discovered feeling of understanding, turned into an indispensable piece of the local area. The ceremonies that were once viewed as dismal turned into an extension between the otherworldly and the everyday, a way for the town to recuperate and develop.

Victor, presently not a main bad guy, arose as an image of flexibility — a man who had confronted the most extreme decisions and, through sympathy and figuring out, tracked down recovery. The town, changed by the drifter's intercession, turned into a microcosm of the more extensive human experience — an update that behind each apparent bad guy, there was a story ready to be heard.

In the last snapshots of this story, as the sun plunged underneath the skyline and the town embraced a fresh start, Victor Belmont stood not as a figure of dread but rather as a demonstration of the groundbreaking force of sympathy. The drifter, playing filled the role of an impetus, proceeded with their excursion, abandoning a town joined by the comprehension that the genuine bad guy was not an outer power but rather the hindrances raised by their own feelings of trepidation and biases.

Thus, the town settled between moving slopes and wandering streams embraced a future formed by the aggregate illustrations learned. The story of Victor Belmont, when a misfortune covered in shadows, turned into an encouraging sign — an update that compassion could unwind the intricacies of the human spirit and that reclamation was feasible in any event, for those marked as bad guys.

The locals, having seen the extraordinary force of understanding, conveyed forward the illustrations got the hang of, guaranteeing that the reverberations of this story waited in the air as a demonstration of the getting through limit with regards to change inside all of us.

6.3 The protagonist faces moral dilemmas and questions the true nature of their power.

In the city of Veridium, a city where transcending high rises came to towards the sky and neon lights painted the night in tints of dynamic tones, carried on with a hero named Evelyn Harper. She was not a standard occupant of Veridium; all things being equal, she had an exceptional gift — or revile, contingent upon one's point of view. Evelyn been able to control time itself, a power that both captivated and tormented her.

Since the beginning, Evelyn found her worldly capacities. From the beginning, it was little and wild — a concise snapshot of history repeating itself or a feeling of pre-cognition. As she became older, notwithstanding, the extent of her powers extended. She could rewind and quick forward time, seeing situation unfurl before her eyes in a nonlinear style. It was a gift that put her aside from the standard residents of Veridium, yet it likewise accompanied a significant weight.

The ethical situations that went with Evelyn's powers started to surface as she understood the possible outcomes of her activities. With the capacity to modify the direction of occasions, she wrestled with inquiries of freedom of thought, predeter-mination, and the moral ramifications of playing with the texture of time. The city, clamoring with life and vast potential outcomes, turned into an intricate embroidery where each choice conveyed weight.

One of Evelyn's earliest upright situations came when she saw an unfortunate mishap unfurling before her. A walker, somewhere out in dreamland, was going to step into the way of an approaching vehicle. Time eased back as Evelyn surveyed the circumstance. The ability to mediate and forestall a disaster lay inside her grip, yet with it came the obligation of changing the normal direction of occasions.

In that crucial second, Evelyn confronted a difficulty that would turn into a common subject in her life — whether to involve her powers for everyone's benefit or stick to the rule of non-obstruction. The idea of playing god, of concluding who lived or passed on, weighed intensely on her inner voice. Eventually, she decided to act, saving the passerby from a heartbreaking destiny.

The far reaching influences of Evelyn's choice became evident in the days that followed. The singular she had saved proceeded to make critical commitments to the local area, changing the existences of people around them. However, as the strings of destiny entwined, Evelyn couldn't shake the pestering uncertainty that her impedance had disturbed the normal request of things.

As Evelyn explored the intricacies of her transient capacities, another ethical situ-ation arose — one that scrutinized the limits of individual connections. The ability to control time offered her brief looks into the eventual fate of her kinships and heartfelt snares. Seeing the expected results, both positive and negative, made a feeling of disquiet.

In a snapshot of weakness, Evelyn ended up enticed to control time to stay away from catastrophe or rescue a faltering kinship. The charm of evading torment and vulnerability pulled at her, provoking a profound reflection into the moral ramifications of utilizing her powers to control the close to home scene of everyone around her. Eventually, she picked restriction, permitting the normal flow of connections to unfurl, regardless of whether it implied confronting despair and dissatisfaction.

The more Evelyn dug into the complexities of her powers, the more she scrutinized the real essence of her gift. Was it a gift or a revile? Did her capacity to control time award her command over her own predetermination, or would she say she was simply a traveler on a foreordained way with the deception of choice?

These existential inquiries drove Evelyn to investigate the philosophical underpinnings of time itself. She dove into antiquated texts, logical speculations, and mysterious precepts, looking for answers that would give lucidity to the mystery of her powers. The more she took in, the more she understood that the real essence of time was a secret that rose above human cognizance.

In the core of Veridium, where the city's heartbeat thump in a state of harmony with the mood of time, Evelyn confronted an ethical junction that would test the actual texture of her being. An approaching fiasco took steps to overwhelm the city — a calamity that could bring about the deficiency of innumerable lives. The impulse to utilize her powers to turn away the approaching misfortune conflicted with the moral quandary of mediating in the normal flow of occasions.

The heaviness of obligation pushed down on Evelyn's shoulders as she wrestled with the choice that lay before her. The city's destiny remained in a critical state, and the ability to modify its fate rested in her grasp. In the tranquil snapshots of consideration, Evelyn addressed whether she reserved the option to play rescuer or whether the city's occupants ought to be permitted to explore their own fates, even despite approaching destruction.

As the seconds ticked away, Evelyn went with a decision — a decision that would challenge the constraints of her powers and power her to stand up to the results of her activities. She decided not to mediate, permitting the normal flow of situation to develop. The disaster struck, and Veridium dove into confusion.

In the fallout, as the city wrestled with the result of the catastrophe, Evelyn scrutinized the profound quality of her choice.

Had she gone with the ideal decision by sticking to the guideline of nonobstruction, or had she deserted her obligation to safeguard the existences of those she had the ability to save? The inner disturbance took steps to unwind the actual groundwork of her convictions.

The city, scarred however tough, started the most common way of remaking. Following the fiasco, Evelyn confronted the investigation of the general population, addressing whether her inaction had been a double-crossing of trust. The ethical key position she had picked turned into a problematic roost, with allegations of carelessness and lack of concern reverberating in the passages of Veridium.

Amidst the public examination, Evelyn looked for comfort in the peaceful corners of the city. The philosophical inquiries that had tormented her brain escalated, and the limits among good and bad obscured into shades of dim. The real essence of her power, when seen as a gift, presently appeared to be an enormous oddity — a power that offered the deception of control while uncovering the intrinsic confusion of presence.

In a snapshot of reflection, Evelyn understood that her powers were not a device for dominance over the universe but rather an unassuming sign of the interconnectedness, everything being equal. The strings of time, fragile and complicated, wove a story that reached out past her singular decisions. The city, individuals, and the situation that unfurled were important for an embroidery that rose above her transient controls.

As Evelyn wrestled with the ethical ramifications of her choices, she looked for counsel from far-fetched sources — an old rationalist who stayed in the core of Veridium. The savant, savvy and baffling, discussed the infinite dance of destiny and choice, encouraging Evelyn to embrace the back and forth movement of the fleeting waterway instead of endeavoring to redirect its course.

The disclosure denoted a defining moment in Evelyn's excursion. The acknowledgment that her powers were not a device for control but rather a focal point through which to observe the excellence and disarray of presence brought a significant feeling of modesty. She started to address the moral utilization of her powers as well as the actual idea of the inquiries she had been posing.

In a city that beat with the musicality of life, Evelyn figured out how to see the value in the minutes that fallen through the fingers of her fleeting controls. The chuckling of youngsters playing in the recreation area, the fragrance of road sellers' delights, and the glow of human association became strong updates that, for all her power, a few parts of presence were outside of her reach.

The ethical situations that had once tortured Evelyn changed into open doors for development and self-disclosure. The city of Veridium, with its clamoring roads and different occupants, turned into a material on which she painted the strokes of her own comprehension. Evelyn embraced the unusual idea of life, understanding that the genuine magnificence of presence lay in its flaws and vulnerabilities.

As Evelyn traveled through the city, her powers turned into a device for perception as opposed to control. She delighted in the accounts unfurling around her — the victories, the misfortunes, and the commonplace minutes that made the orchestra out of Veridium. The ethical inquiries that had once tortured her currently assumed a lower priority in relation to the enthusiasm for the present.

In the last snapshots of this story, Evelyn remained on the slope of a high rise, sitting above the city she had come to grasp in manners that rose above fleeting control. The neon lights gleamed underneath, projecting a kaleidoscope of varieties in the city. The city beat with life, and Evelyn, when a hero scrutinizing the real essence of her power, tracked down comfort in the acknowledgment of the enormous dance that bound every living thing.

As the night embraced Veridium, Evelyn Harper, with a newly discovered feeling of lowliness, proceeded with her excursion through the embroidery of time. The inquiries that had once tormented her became reverberations in the breeze, tokens of an extraordinary odyssey that had disentangled the secrets of ethical quality, pre-determination, and the fragile harmony between the ability to notice and the insight to give up. The city, with its constant cadence, conveyed forward the reverberations of Evelyn's excursion — a demonstration of the persevering through intricacy of presence and the significant examples learned in the dance between the past, present, and future.

Chapter 7

Absolute Zero Hour

In the tremendous field of the universe, where the endless ranges of reality combine, there exists a place of outright quietness, a second frozen in the constant dance of presence. This is Indisputably the Party time, an idea that rises above the limits of traditional comprehension and dives into the profundities of supernatural thought.

In the domain of physical science, outright zero addresses the most reduced conceivable temperature, the place where particles stop all movement. It is a condition of wonderful request, a second frozen in the timeless hug of nothingness. This hypothetical temperature, frequently represented by the image 0 K, is a deliberation that challenges the actual pith of our cognizance. At this temperature, the laws of thermodynamics direct that entropy, the proportion of turmoil in a framework, arrives at its base worth. It is a mark of outright quiet, where disarray gives way to peacefulness.

In any case, Unquestionably the Party time isn't simply an actual idea; it reaches out past the bounds of logical talk. It is a representation for a significant existential second, a crossroads where the flows of time meet and reality itself is suspended. It is a place of change, a limit between what and will be.

As one thinks about the mystery of Irrefutably the Party time, it becomes apparent that its importance goes past the substantial world. It is an image of change, a second pregnant with potential. In the vast embroidery of presence, this hour denotes a delay, an accentuation in the story of the universe.

Envision an excursion through the universe, navigating the heavenly scenes that stretch across the texture of room. As you approach Without a doubt the Party time, the actual texture of reality starts to twist and turn. Time itself turns into a pliable substance, bowing and distorting in manners that challenge coherent clarification. It is an inestimable expressive dance, a dance of particles and waves, prompting a definitive crescendo — Without a doubt the Party time.

Right now, all that was once moving comes to a standstill. Stars, worlds, and grandiose residue are suspended in an enormous scene, frozen in a fragile dance that opposes the determined walk of time. A scene rises above the restrictions of human creative mind, a brilliant movement organized by the concealed hand of the universe.

As one stands at the edge of Unquestionably the Party time, a significant feeling of stunningness and miracle grabs hold. The actual idea of presence is exposed, deprived of the deceptions that shroud the universe in secret. In this frozen moment, the mysteries of the universe are murmured to the people who set out to tune in.

The excursion to Unquestionably the Party time isn't for weak willed. It requires a readiness to defy the obscure, to wrestle with the key inquiries that have reverberated through the halls of human idea for centuries. What lies past the cover of our getting it? What secrets anticipate the people who try to look into the chasm?

The mission for Unquestionably the Party time is, fundamentally, a journey for importance. It is an excursion into the profundities of presence, a quest for the slippery bits of insight that evade the grip of the limited brain. A journey rises above the limits of science and reasoning, digging into the actual quintessence of being.

As the valiant wayfarer moves toward Without a doubt the Party time, the limits among self and universe start to obscure. The differentiation among spectator and noticed disintegrates, and a significant feeling of interconnectedness swarms the cognizance. Right now, the universe isn't just an object of study; it is a no nonsense substance with which the spectator is personally weaved.

The idea of outright zero is established in the laws of thermodynamics, the core values that administer the way of behaving of energy and matter. At this hypothetical temperature, particles grind to a halt, and entropy arrives at its base worth. It is a mark of harmony, a condition of wonderful request that challenges how we might interpret the universe.

Chasing after Unquestionably the Party time, researchers and thinkers the same wrestle with the ramifications of this subtle idea. Is it a hypothetical cutoff that can never be reached, or does it address an unmistakable objective in the tremendous scope of the universe? The responses lie covered in the secrets that wrap the actual texture of the real world.

In the records of logical request, the mission for outright zero has prompted historic disclosures and mechanical headways. Cryogenics, the part of physical science that arrangements with the creation and impacts of extremely low temperatures, has opened new outskirts in medication, materials science, and then some. The capacity to control temperatures with accuracy has opened ways to domains of probability that were once considered inaccessible.

However, the excursion to Indisputably the Party time isn't bound to the labs of physicists. An excursion unfurls in the openings of the human brain, a mission for understanding that rises above the limits of experimental perception. The traveler of this enormous outskirts isn't furnished with instruments and conditions alone; they use the apparatuses of reflection and consideration.

As the wayfarer digs further into the secrets of Unquestionably the Party time, they experience the mysteries that lie at the core of presence. The actual idea of a temperature at which particles stop all movement challenges how we might interpret reality. An idea challenges instinct, driving the searcher into a domain where the recognizable laws of the regular world never again hold influence.

Chasing this subtle second, the voyager wrestles with the idea of time itself. Unquestionably the Party time isn't simply a point in the transient scene; it is an entryway to the boundless. As the pioneer moves toward this edge, the regular progression of time starts to unwind, uncovering an embroidery of interconnected minutes that stretch out past the direct movement of past, present, and future.

The mission for Unquestionably the Party time is an excursion of reflection and disclosure. It is a journey into the profundities of cognizance, where the wayfarer faces the key inquiries that have tormented humankind starting from the beginning of mindfulness. What is the idea of presence? What is the reason that drives us forward into the unfamiliar regions of the universe?

As the pilgrim wrestles with these existential inquiries, Unquestionably the Party time takes on an emblematic importance. It turns into a similitude for the significant snapshots of tranquility that intersperse the human experience. Amidst life's tumultuous dance, there are epiphanies and disclosure, where the clamor of the world disappears, and the searcher is left with a significant feeling of presence.

The idea of outright zero, with its suggestions for the discontinuance of movement, reverberates with the human longing for tranquility in the midst of the choppiness of life. Chasing after progress, love, and satisfaction, people frequently end up trapped in the steady flows of aspiration and want. Irrefutably the Party time turns into a guide, an update that in the midst of the turmoil, there exists a mark of balance where the spirit can track down relief.

In the domain of otherworldliness, Unquestionably the Party time takes on a mysterious importance. It is a snapshot of greatness, where the limits between oneself and the heavenly haze. The searcher, as they continued looking for significance and reason, moves toward this hallowed edge with love and amazement. It is an excursion into the numinous, a journey that rises above the impediments of the material world.

As the adventurer explores the unknown regions of Irrefutably the Party time, they stand up to the shadows that prowl in the openings of the human mind. The frozen tranquility of this grandiose second turns into a mirror, mirroring the deepest feelings of trepidation and wants of the searcher. It is a showdown with oneself, an investigation of the profundities that lie underneath the outer layer of cognizant mindfulness.

The excursion to Irrefutably the Party time isn't without its difficulties. The pilgrim experiences snags that test the constraints of their purpose and understanding. The actual idea of the mission opposes simple order, and the searcher should wrestle with vulnerability and uncertainty. It is an excursion into the obscure, an act of pure trust into the vast void.

Chasing after this subtle second, the pioneer is stood up to with the constraints of language and reason. Without a doubt the Party time rises above the limits of traditional talk, escaping exact definition and depiction. An idea coaxes the searcher to go past the bounds of words and embrace the secrets that lie past the shroud of language.

As the pilgrim remains at the limit of Irrefutably the Party time, they are stood up to with the immeasurability of the universe and the vastness of potential outcomes that stretch before them. It is a snapshot of significant decision, a point where the direction of the excursion is molded by the choices of the searcher. Will they keep on wandering into the grandiose obscure, or will they retreat to the wellbeing of the natural?

In the repercussions of the excursion to Without a doubt the Party time, the pioneer arises changed. The experiences acquired from the grandiose journey reverberate in the profundities of their being, molding the manner in which they see the world. The frozen quietness of the grandiose second turns into a standard — a reference point that grounds the searcher in snapshots of tumult and vulnerability.

Irrefutably the Party time, with its suggestions for the discontinuance of movement, is an idea that rises above the limits of logical request and philosophical examination. It is an excursion into the core of presence, a journey for significance and understanding that goes past the constraints of language and reason. The wayfarer, equipped with interest and boldness, sets out on an infinite odyssey that challenges the actual texture of the real world.

In the excellent embroidery of the universe, Without a doubt the Party time is an enormous accentuation — a snapshot of tranquility that reverberations through the hallways of time. It is an update that, in the midst of the perpetual dance of particles and waves, there exists a place of balance where the secrets of the universe are revealed. The traveler, having navigated the huge scope of the astronomical scene, remains at the edge of understanding, prepared to embrace the consistently unfurling secrets that lie past.

7.1 Reach a climactic point where the group confronts the mastermind behind the temperature conspiracy.

In the shadowy maze of scheme and interest, where the strings of trickiness weave an embroidery of vulnerability, the gathering of bold people had been on a steady mission to disentangle the secret behind the control of temperatures. It was a trick that rose above the limits of science and dove into the profundities of evil thought processes. As they explored the exciting bends in the road of the puzzler, their process arrived at a climactic point — one where the subtle pretense was ready to be lifted, uncovering the genius coordinating the temperature trick.

The gathering, a different gathering of researchers, examiners, and masterminds, had sorted out sections of proof that indicated a stupendous plan to control the

World's environment. A lot was on the line, for the actual equilibrium of nature remained in a critical state. Unexplained irregularities in temperature designs, outrageous climate occasions, and a developing feeling of disquiet had powered their quest for reality. The maze had tried their purpose, however it had likewise fashioned a strong bond among them — an aggregate assurance to defy the powers controlling the actual texture of the planet.

Their process had driven them through the passages of force and the underside of furtive associations. They had translated mysterious messages, disentangled coded correspondences, and outfoxed the people who tried to keep the connivance hidden in mystery. It was a waiting game, with each disclosure carrying them nearer to the core of the secret.

The gathering's journey veered off in a strange direction when they uncovered a surreptitious lab settled in a remote corner of the world. It was a cutting edge office, stowed away from intrusive eyes, where trials to control worldwide temperatures were being directed. As they penetrated the office, the size of the connivance became clear. Trend setting innovation, taken research, and an odious plan highlighted a genius coordinating an arrangement of worldwide extents.

Profound inside the guts of the secret research facility, the gathering found a chamber that housed the controls for a gadget fit for changing environmental circumstances on a planetary scale. It was a mechanical wonder, a dismal sign of the driving force's vile desires. The gathering knew that facing the orchestrator of this temperature trick was basic, for their own endurance as well as for the eventual fate of the whole planet.

The showdown unfurled in a faintly lit control room, where screens flashed with information and unpropitious murmurs exuded from the complicated hardware. As the gathering mindfully progressed, their faculties uplifted, a figure rose up out of the shadows — the slippery brains behind the temperature intrigue.

The brains, clad in a shroud of egotism and prevalence, reviewed the gathering with a chilling look. A grin played all the rage, a determined articulation that double-crossed a brain familiar with remaining strides in front of the people who looked to challenge their maneuvers. The air in the room snapped with strain, the air weighty with the heaviness of approaching disclosure.

"You've overcome much, my fearless agents," the driving force proclaimed, their voice dribbling with haughtiness. "However, you are past the point of no return. The wheels of my arrangement are as of now moving, and there's no way to stop it."

The gathering, unfazed by the driving force's grandiosity, stood fearless. Their chief, a carefully prepared specialist with a steely look, ventured forward and stood up to the puppeteer organizing the temperature intrigue.

"We understand what you're doing," the pioneer stated, their voice enduring. "Your control of worldwide temperatures, the devastating occasions you've released — it closes here. The world has the right to know reality."

The genius laughed, a sound that resonated through the control room. "The fact of the matter is a relative idea, my dear examiner. What you see as control, I see as essential adjustment. The world is near the very edge of turmoil, and just through my mediation can adjust be reestablished."

The gathering traded vigilant looks, detecting that the brains' thought processes were covered in a wound reasoning that supported their activities. The showdown was not only a conflict of philosophies; it was a fight between the protectors of normal request and a power looking to declare territory over the very components that supported life.

As the verbal fighting escalated, the driving force uncovered the genuine degree of their arrangement. They discussed a world wavering on the edge of natural breakdown, a planet moved as far as possible by the overabundances of human development.

In their distorted point of view, the control of temperatures was a frantic endeavor to reset the equilibrium, to guide mankind away from the cliff of implosion.

The gathering tuned in, doubtful yet enamored by the genius' hoodwinked vision of salvation. It became clear that the scheme ran further than a simple mission for power or control. The brains saw themselves as a rescuer, a harbinger of progress in a world tilting toward its own death.

As the genius elucidated their stupendous arrangement, the gathering immediately jumping all over the chance to accumulate significant data. They secretly got to the research center's centralized server, downloading documents that itemized the whole extent of the temperature control project. The information uncovered the degree of the ecological disasters designed by the driving force, as well as their mind boggling trap of impact that stretched out into states, companies, and logical establishments.

Furnished with this dooming proof, the gathering defied the driving force with reality. The screens in the control room gleamed to life, showing undeniable proof of the natural devastation unleashed upon the planet. The driving force's demeanor faltered, a flashing break in their veneer of unflinching certainty.

"You can't stifle reality any longer," the examiner pronounced, waving the proof like a weapon. "Your arrangement has caused hopeless harm, and the world will consider you responsible."

The driving force, cornered and uncovered, confronted a retribution. The gathering, filled by a feeling of equity and the direness of the worldwide emergency, squeezed for replies. They requested to know the genuine thought processes behind the control of temperatures, trying to grasp the wound reasoning that had driven the genius to such extreme measures.

In the strained trade that followed, the brains' facade of pomposity started to disintegrate. Underneath the exterior of a self-declared hero, a more evil truth arose. It was a disclosure that rose above the domain of natural control and dug into the profundities of individual grudges and uncontrolled desire.

The brains, it ended up, held onto a firmly established resentment against the world — a hatred brought into the world from saw treacheries and disloyalties. Their activities were not just an off track endeavor to save the planet yet a vindictive plot to achieve bedlam and experiencing on a worldwide scale. The temperature intrigue, with all its devastating outcomes, was an indication of an injured mind looking for revenge.

As the gathering consumed this disclosure, a significant need to keep moving grasped them. The brains' arrangement was still moving, and the gathering comprehended that they needed to act quickly to forestall further ecological obliteration.

With the proof close by, they reached specialists, states, and natural associations, sharing reality with regards to the genius' slippery plot.

The showdown with the genius had arrived at a peak, yet the genuine fight lay in forestalling the disastrous occasions put into high gear by their malicious arrangement. The gathering, presently furnished with partners and a worldwide organization of help, left on a test of skill and endurance to impede the looming ecological calamities.

The driving force, understanding that their amazing plan was disentangling, endeavored to get away from the grip of equity. A high-stakes pursue followed, traversing mainlands and testing the gathering's strength. The pursuit drove them through remote scenes, clamoring urban communities, and tricky territories. Each step carried them closer to capturing the brains and ending the devastating chain of occasions that undermined the planet.

In the last standoff, against the setting of a climactic union of powers, the gathering stood up to the driving force in a ruined scene. The genius, frantic and cornered, released a blast of way of talking, endeavoring to legitimize their activities as a means to an end. Yet, the gathering, braced by the information on the genius' actual thought processes, stood enduring in their obligation to equity.

As the showdown arrived at its pinnacle, a tempest assembled above, reflecting the whirlwind inside the hearts of those present. Thunder thundered, and lightning broke across the sky, an essential showcase repeating the vast stakes of the fight. The genius, understanding the purposelessness of their endeavors, surrendered to the inescapable.

In a snapshot of fitting retribution, the genius was secured, their rule of natural dread finished off. As they were driven away in guardianship, the gathering watched with a blend of fulfillment and exhaustion. The fight had been won, however the scars of the trick waited as natural calamities that necessary quick consideration.

In the repercussions of the climactic showdown, the gathering worked resolutely to alleviate the effect of the driving force's controls. They teamed up with researchers, legislatures, and natural associations to execute measures that would reestablish biological equilibrium and mend the injuries incurred upon the planet. It was an overwhelming undertaking, however the gathering, presently joined by a common perspective, confronted the difficulties with resolute assurance.

The disclosure of the temperature trick sent shockwaves through the world. General society, at first ignorant about the coordinated ecological debacles, wrestled with the extent of the disclosures. States sent off examinations, considering those complicit in the connivance responsible for their activities. The worldwide local area, stirred to the delicacy of the planet, energized together to address the pressing ecological emergencies.

As the gathering pondered their excursion, they comprehended that the fight against the genius was only one part in a continuous battle to safeguard the Earth. The temperature scheme had been uncovered, however the more extensive difficulties of environmental change and natural debasement continued. The gathering, presently perceived for their chivalrous endeavors, became advocates for maintainable practices, ecological mindfulness, and the conservation of the planet.

The climactic showdown with the driving force had impeded a pernicious plot as well as touched off a flash of shared perspective. The world, joined by the common danger to its actual presence, started to rethink the connection among humankind and the climate. It was a defining moment — a snapshot of retribution that prodded a worldwide development toward mindful stewardship of the planet.

In the repercussions of the temperature connivance, the gathering arose as unrecognized yet truly great individuals, their excursion a demonstration of the flexibility of the human soul. The climactic showdown had not just uncovered the intrigues of a turned brain however had likewise enlivened an aggregate reaction to shield the Earth for people in the future. The brains' endeavor to control temperatures had coincidentally set off a worldwide arousing, and the gathering, having confronted the void of ecological disaster, played had an essential impact in directing mankind toward an additional feasible and amicable future.

7.2 Uncover the true potential and limitations of the protagonist's power.

In the core of a world permeated with the phenomenal, the hero remained at the nexus of force and vulnerability. Gifted with an extraordinary capacity that opposed the regular limits of human potential, they wrestled with the conundrum of their own reality. As the story unfurled, the hero left on an excursion of self-disclosure, driven by the longing to reveal the genuine degree of their power and figure out the restrictions that bound them.

From the start, the hero's power appeared to be boundless — a power that could reshape reality itself. With a simple idea, they could control the components, navigate the texture of room, and even look into the embroidery representing things to come. It was a gift that put them aside, an enormous irregularity that welcomed wonder and esteem. Nonetheless, underneath the outer layer of this unprecedented capacity hid a maze of intricacies and questions.

The excursion to uncover the genuine capability of the hero's power started with provisional trial and error. They tried the limits of their capacities, pushing against the limitations of the known and wandering into strange domains of the phenomenal.

The world turned into their lab, and each thought, each expectation, conveyed the heaviness of probability.

As the hero dove further into the investigation of their power, they found that its actual potential was unpredictably associated with the scene of their feelings and expectations. The profound inclinations that streamed inside them turned into the impetus for the indication of their capacities. Satisfaction and love birthed amicable indications, while outrage and dread undulated into turbulent mutilations of the real world.

The disclosure of this close to home beneficial interaction was both engaging and overwhelming for the hero. It implied that the material of their power was painted with the tints of their deepest sentiments, and dominating command over their feelings became vital. The hero before long understood that their own mind was a landmark, where the rhythmic movement of feelings directed the actual idea of their capacities.

With this recently discovered understanding, the hero started a trained excursion of contemplation and close to home dominance. They looked for the direction of guides who had crossed comparable ways, diving into antiquated shrewdness and obscure lessons that offered bits of knowledge into the sensitive equilibrium expected to mindfully employ such unprecedented power. The hero's development reflected a profound odyssey — a journey for edification that rose above the limits of the ordinary.

As the hero sharpened their profound control, the genuine capability of their power started to appear in stunning showcases of authority. They could invoke deceptions that moved like ethereal ghosts, control time to witness the reverberations of the past and impression the conceivable outcomes representing things to come, and order the components with an artfulness that reflected the ensemble of creation itself. The world turned into a material whereupon the hero painted the unprecedented, a demonstration of the vast possible that lay inside.

However, with each step toward authority, the hero experienced the fragile dance among power and obligation. The outcomes of uncontrolled feelings lingered like shadows, prepared to mutilate the actual texture of the real world. The hero wrestled with moral predicaments, the heaviness of their decisions amplified by the repercussions that reverberated through the woven artwork of presence.

The impediments of the hero's power additionally became evident as they explored the complexities of their capacities. The limits between the phenomenal and the unremarkable were not generally obvious, and the hero observed that their power was not a panacea for every one of the difficulties they confronted. It couldn't repair broken connections, delete the scars of the past, or safeguard them from the inescapable preliminaries of life.

The limits of the hero's power appeared surprisingly — minutes when the actual idea of their capacities turned into a wellspring of seclusion. The gift that put them aside additionally raised an undetectable hindrance among them and the remainder of mankind.

The weight of understanding the intricacies of the real world, the heaviness of seeing the strings of destiny, and the segregation that accompanied the uncommon set the hero untied in an ocean of isolation.

This separation was not only physical yet additionally existential. The hero wrestled with the existential forlornness that went with the consciousness of the delicacy of human life. They could see the fragile dance of life, the transient magnificence of each and every second, and the certainty of mortality. This elevated mindfulness cast a shadow over their cooperations, causing the embroidery of human association with appear to be both fleeting and valuable.

As the hero explored the maze of their own limits, they looked for comfort in the bonds they framed with the people who could see the exceptional idea of their excursion. The coach who had directed them, the friend who partook in their battles, and the buddies who remained by them became mainstays of help despite existential difficulties. It was through these associations that the hero tracked down strength, an update that even the uncommon required the anchor of human association.

The hero's power likewise bore the heaviness of an ethical obligation that rose above private cravings. The capacity to control reality conveyed with it the moral basic to think about the results of each and every idea and activity. The hero wrestled with the quandary of through and through freedom versus determinism, scrutinizing the ramifications of their power on the independence of those whose lives they contacted.

This ethical dilemma arrived at a climactic moment that the hero confronted a decision that could change the direction of predetermination itself. The strings of destiny combined, and the hero remained at the junction of a choice that would echo through the embroidery of presence. The weight of such a decision was significant, and the hero wound up torn between the craving to shape a considerate reality and the mindfulness that their intercessions could disturb the fragile equilibrium of the universe.

In the cauldron of this ethical retribution, the hero defied the real essence of their power — a power that could be both an encouraging sign and a harbinger of turmoil. The obligation that accompanied such uncommon capacities weighed intensely on their shoulders, and the excursion to comprehend the restrictions of their power turned into a strong investigation of the convergence between freedom of thought and fate.

As the hero wrestled with these significant inquiries, they found that the genuine capability of their power lay in the control of reality as well as in the extraordinary effect of their decisions. The capacity to shape fates, to wind around the strings of presence with cognizant goal, turned into a demonstration of the ethical compass that directed their activities.

The hero's excursion to uncover the genuine potential and restrictions of their power turned into an embroidery of self-disclosure, moral arousing, and existential consideration. The unprecedented capacities that put them aside turned into a pot through which they fashioned a more profound comprehension of their own mankind. The impediments of their power, a long way from being requirements, turned into the impetus for development, flexibility, and a significant appreciation for the excellence inborn in the human experience.

In the end result of their excursion, the hero tracked down harmony — a fragile harmony between the exceptional and the commonplace, the power and the restrictions, the disengagement and the association. The embroidery of their reality bore the engravings of hardships, however it likewise reverberated with the agreeable harmonies of self-acknowledgment and a significant feeling of direction.

The hero, having explored the maze of their own capacities, arose not as an exceptional figure but rather as a modest watchman of the remarkable. Their power, when an infinite riddle, turned into a device for sympathy, compassion, and positive change. The hero, having revealed the genuine potential and impediments of their power, embraced the inborn obligation that accompanied the uncommon, winding around an inheritance that rose above the bounds of simple presence — a heritage scratched into the actual texture of the real world.

7.3 Explore themes of sacrifice, loyalty, and the consequences of wielding such a unique ability.

In the huge embroidery of human experience, where the strings of destiny entwine with our decisions, subjects of penance, steadfastness, and the outcomes of employing exceptional capacities weave a story that rises above the normal. The story unfurls in reality as we know it where the exceptional isn't simply a chance yet a reality, and the hero, favored or troubled with a power that opposes the standard, ends up ensnared in the perplexing dance of these significant subjects.

Penance, a topic carved in the actual texture of mankind's set of experiences, arises as a piercing theme in the hero's excursion. As they explore the maze of their special capacity, they are stood up to with decisions that request significant penances. The idea of their power, a power that can reshape reality itself, requires a giving up of individual longings and an acknowledgment of the weights that accompany such exceptional gifts.

Right off the bat in the hero's excursion, they experience a significant second where the prosperity of a friend or family member remains in a precarious situation. The ethical problem gives itself an unmistakable clearness — their interesting skill could be utilized to modify the direction of occasions and save a day to day existence, however at what cost? The penance required isn't just a physical or material one; it is an acquiescence of the hero's own close to home prosperity, a demonstration of the cost that employing such power removes.

At this time of penance, the hero wrestles with the moral ramifications of assuming the part of a grandiose judge. The strings of destiny, when liquid and unusual, presently appear to be dependent upon their order. The heaviness of obligation pushes ahead upon them, and the penance they make for a friend or family member turns into a defining moment in how they might interpret the results of using exceptional capacities.

Faithfulness, a subject entwined with penance, turns into a core value for the hero. As they cross the complex scenes of force and obligation, they structure bonds with people who see past the cloak of the exceptional. These associations — manufactured in the cauldron of shared encounters, trust, and common getting it — become the bedrock of the hero's excursion.

The dependability of mates who stand by the hero's side, notwithstanding the wild idea of their capacities, turns into a wellspring of solidarity and comfort. These people, each wrestling with their own preliminaries, represent a fellow devotion fashioned in the pot of difficulty. The subject of reliability isn't restricted to simple brotherhood; it turns into a demonstration of the strength of human association even with the exceptional.

As the hero's process unfurls, faithfulness is tried in the pot of affliction. Double-crossings, stowed away thought processes, and the consistently present phantom of individual plans cast a shadow over the bonds that have been framed. The steadfastness of friends turns into a sensitive dance, a harmony among trust and incredulity. The hero, outfitted with the capacity to see the secret flows underneath the surface, explores the complicated territory of dependability with an insightful eye.

The results of using such an exceptional capacity reach out past the domain of a disregard for one's own needs and steadfastness. The actual texture of the truth is molded by the hero's goals, and each figured helps the possibility to echo through the embroidery of presence. The results are not generally unsurprising, and the hero wrestles with the moral consequences of assuming the part of a grandiose draftsman.

In an essential second, the hero faces an ethical predicament that tests the actual groundwork of their dependability. A decision looms not too far off — one that includes forfeiting the prosperity of an ally for everyone's best interests. Unwaveringness to a bigger reason conflicts with the singular associations that have been fashioned, and the hero is compelled to stand up to the inborn strain between private bonds and the infinite obligation that accompanies their novel capacity.

The outcomes of this decision resonate through the account, making a permanent imprint on the connections that characterize the hero's excursion. The subject of penance, when based on private decisions, presently stretches out to the penance of standards, trust, and the actual embodiment of being human.

The results become a mirror, mirroring the's comprehension hero might interpret the fragile harmony between using power and exploring the complexities of the human experience.

As the story unfurls, the outcomes of the hero's activities become progressively intricate. The modifications they make to the real world, when seen as generous intercessions, uncover unexpected expanding influences. The actual texture of presence, fragile and interconnected, opposes control, and potentially negative side-effects manifest in manners that challenge the's comprehension hero might interpret circumstances and logical results.

The results of using such an exceptional capacity likewise reach out to the moral and existential components of the hero's excursion. The ability to shape reality, when a wellspring of strengthening, turns into a weight that weighs intensely on their shoulders. The moral pickles that emerge from assuming the part of a vast judge force the hero to defy inquiries of freedom of thought, predetermination, and the ethical obligation that goes with their remarkable capacities.

In a climactic second, the results of the hero's activities arrive at a crescendo. The actual texture of reality starts to disentangle, contorted by the potentially negative results of their intercessions. The results stretch out past the individual domain, influencing the existences of the people who exist on the fringe of the hero's excursion. The topic of penance, unwaveringness, and results meets in a frenzy of grandiose extents, testing the actual furthest reaches of the hero's power and the strength of their personality.

As the hero wrestles with the outcomes of their activities, they go through an extraordinary cycle. The acknowledgment that using such an extraordinary capacity conveys individual outcomes as well as infinite ramifications turns into a pot for self-disclosure. The hero, when characterized by the phenomenal idea of their power, arises as a more nuanced and mindful individual, insightful of the fragile equilibrium expected to explore the perplexing dance of penance, devotion, and results.

The goal of the story is certainly not a slick unwinding of strings yet a mosaic of intricacies. The results of the hero's decisions make a permanent imprint on the world they possess, molding the fates of those contacted by their phenomenal excursion. The topic of penance, when an individual weight, turns into an aggregate story — a common inheritance that rises above the limits of individual experience.

In the conclusion, the hero thinks about the excursion, wrestling with the significant topics that have characterized their odyssey. Penance, when seen as a lone demonstration, is currently seen from the perspective of aggregate liability. The faithfulness manufactured in the cauldron of misfortune turns into a demonstration of the strength of human association despite the remarkable. The results of employing such an exceptional capacity, a long way from being a weight to bear alone, become a common story — an embroidery woven from the strings of penance, steadfastness, and the unyielding soul that characterizes the human experience.

In the domain where the conventional and the unprecedented meet, the outcomes of using a remarkable capacity echo through the texture of the real world. The hero, troubled or favored with a power that opposes the standards of presence, wrestles with the heaviness of their decisions and the unanticipated repercussions that reverberation through the embroidery of their excursion.

The idea of the hero's one of a kind capacity is both a wellspring of strengthening and an impetus for significant outcomes. It permits them to shape reality itself, to navigate the limits of what is seen as could be expected. However, this power is a two sided deal, slicing through the sensitive strings of circumstances and logical results with outcomes that stretch out past the prompt.

As the hero explores different avenues regarding their capacity, the results of their activities become obvious. Each thought, each expectation, each control of the remarkable accompanies a cost. The main notion of this reality happens when an apparently harmless modification to reality prompts unexpected changes in the existences of people around them.

The results are not prompt or direct all the time. They unfurl like waves on a lake, contacting far off shores in manners that the hero could never have expected. An opportunity choice to change a second in time might set off a chain response of occasions, modifying predeterminations and entwining lives in manners that make no sense. The hero, when inebriated by the capability of their power, presently faces the sobering reality that each decision worries about a concern of liability.

The topic of outcomes escalates as the hero wrestles with moral difficulties. The ability to shape reality, while charming, requests an insightful moral compass. decisions in the pot of uncommon power become a demonstration of the hero's personality, uncovering the real essence of their aims and the rules that guide their activities.

An essential second emerges when the hero is confronted with a decision that has broad outcomes. The strings of destiny combine, and the hero remains at an intersection where the outcomes of their choice could modify the direction of fates. The heaviness of such a decision is obvious, and the hero, outfitted with the capacity to see the outcomes before they unfurl, should wrestle with the ethical basic that accompanies their remarkable power.

The results of this critical decision reverberate all through the story, molding the connections and elements that characterize the hero's excursion. The subject of penance interlaces with the outcomes, as the hero understands that each choice to modify reality accompanies a penance — be it individual, moral, or existential. The results of their activities become indistinguishable from the penances they make for the sake of a more prominent reason or an individual conviction.

The results additionally stretch out to the hero's connections. The power they use makes a split among them and the individuals who can't fathom the complexities of controlling reality. Companions become careful, partners question thought processes, and the outcomes of being a phenomenal person in a world limited by the conventional become clear. The subject of detachment, brought into the world from the results of employing such a remarkable capacity, turns into an impactful suggestion in the hero's excursion.

As the hero wrestles with the steadily growing results of their activities, they are gone up against with the delicacy of human life. The ability to shape the truth isn't a panacea; it can't patch broken hearts, delete the scars of the past, or safeguard friends and family from the inescapable preliminaries of life. The results of employing such a special capacity become a lowering acknowledgment that there are parts of presence unchangeable as far as the hero might be concerned.

The moral ramifications of their power escalate as the outcomes of their activities become progressively complicated. The hero, when certain about their capacity to perceive right from wrong, is compelled to defy the ill defined situations of profound quality. The results of mediations that appear to be kindhearted on a superficial level uncover unexpected moral problems, moving the hero to scrutinize the actual underpinnings of their standards.

The results likewise stretch out to the hero's own mind. The weight of molding reality, of bearing the obligation regarding the decisions made, negatively affects their psychological and close to home prosperity. The outcomes of their power become an existential weight, inciting the hero to scrutinize the reason and significance of their excursion. The subject of self-revelation combines with the outcomes, as the hero digs into the profundities of their own spirit looking for understanding.

In a climactic second, the outcomes of the hero's activities arrive at a peak. Reality itself appears to defy the controls, and the fragile harmony between the conventional and the remarkable is disturbed. The outcomes manifest in a bedlam of vast extents, testing the actual furthest reaches of the hero's power and the versatility of their personality.

As the outcomes unfurl, the hero is compelled to stand up to the constraints of their capacity. The power that once appeared to be endless now uncovers its weaknesses. The results become a mirror, mirroring the complex dance among power and obligation. The hero, lowered by the results of their activities, should wrestle with the acknowledgment that even the phenomenal has its limits.

The topic of results interweaves with the idea of accidental results. The hero, notwithstanding their capacity to see the possible results prior to settling on a decision, finds that the complexities of the truth are excessively complicated to be completely expected.

The outcomes, as boisterous strings, weave a story that opposes the hero's endeavors to control each part of their excursion.

In the outcome of the climactic results, the hero is left to deal with the aftermath. The world they occupy bears the characteristics of their decisions, and the results become an aggregate story — a common inheritance that rises above the bounds of individual experience. The hero, presently receptive to the fragile harmony among power and outcome, should explore the repercussions with a recently discovered comprehension of the obligations that accompany their novel capacity.

The outcome turns into an intelligent second for the hero. The outcomes of their excursion, both individual and enormous, become an embroidery woven from the strings of penance, steadfastness, and the unyielding soul that characterizes the human experience. The topic of results, when an overwhelming power, changes into a wellspring of insight — an aide for the hero as they explore the steadily moving scene of the unprecedented.

The hero rises up out of the pot of results changed. The excursion, with every one of its exciting bends in the road, turns into a demonstration of the flexibility of the human soul notwithstanding the unprecedented. The outcomes, when seen as a weight, become a wellspring of development, self-revelation, and a significant appreciation for the mind boggling dance of circumstances and logical results that characterizes the human experience. The hero, having crossed the perplexing embroidery of results, remains at the edge of another comprehension — a savvier, more nuanced individual, mindful of the fragile harmony between using power and exploring the unpredictable strings of predetermination.

Chapter 8

Equilibrium Epiphany

In a world wavering near the precarious edge of disorder, where the fragile equilibrium of presence balanced by the most slender string of harmony, there arose a person whose very presence would turn into the impetus for a significant revelation. This individual, honest for all intents and purposes however holding onto a brain that reverberated with the reverberations of undiscovered insight, would set out on an excursion that rose above the limits of the known and wandered into the domains of the remarkable.

The setting was a rambling city, a microcosm of the battles and goals of mankind. High rises extended towards the sky, their intelligent surfaces reflecting the aggregate dreams and wants of the people who possessed the clamoring city. However, underneath the façade of progress and success, a propensity of agitation beat through the roads. Society, similar to a pendulum, swung among request and turmoil, each passing second taking steps to tip the fragile harmony in one bearing or the other.

Amidst this metropolitan embroidery, our hero arose — a figure covered in secret, traveling through the packed roads with a reason that rose above the everyday. With each step, the individual appeared to assimilate the energy of the city, adjusting themselves to the beat of the city. Unbeknownst to the individuals who brushed shoulders with this confounding drifter, a change was in progress — an enlivening that would modify the actual texture of their discernment.

The hero's name, on the off chance that such a traditional mark could be applied, was unimportant in the excellent plan of their reality. They were a vessel for a higher comprehension, a channel through which the universe tried to speak with itself. The excursion of Harmony Revelation had started.

As the hero dug further into the core of the city, the differentiations between the different features of human experience turned out to be progressively evident. On one road, extravagant retail facades showed the most recent mechanical wonders, promising a future formed by development and progress. On another, decrepit structures remained as quiet observers to the battles of those underestimated by the persevering walk of time. The juxtaposition of riches and neediness, of honor and hardship, painted a striking scene of the cultural lopsided characteristics that took steps to disturb the harmony.

It was inside this embroidery of inconsistencies that the hero experienced a different cast of characters, each addressing an interesting feature of the human experience. There was the disappointed corporate chief, shackled by the requests of a vicious world that deliberate outcome in net revenues. Across the road, a road craftsman communicated insubordination through lively wall paintings, rocking the boat with strokes of inventiveness.

In the core of the city, where the commotion of day to day existence arrived at its crescendo, the hero coincidentally found an undercover get-together — an aggregate of scholars, logicians, and visionaries who looked to disentangle the secrets of presence. This diverse gathering, drawn together by an imperceptible power that rose above friendly and social limits, turned into the cauldron for the hero's prospering revelation.

As the hero took part in discussions with these lights, an embroidery of thoughts unfurled — a rich mosaic that crossed the range of human idea. Conversations went from the idea of reality to the meaning of individual awareness in molding the aggregate fate of mankind. It was here, in the midst of the free trade of thoughts, that the hero encountered a significant disclosure — a revelation that broke the constraints of regular getting it.

The substance of this disclosure lay in the interconnectedness, everything being equal. The hero saw the strings that bound the dissimilar components of presence into a strong entire — an enormous dance wherein each thought, each activity,

undulated across the texture of the real world. It was an acknowledgment that rose above the dualities of good and detestable, good and bad, and embraced the inborn solidarity of the universe.

Equipped with this recently discovered understanding, the hero set off to challenge the powers that took steps to upset the sensitive harmony of the city. Helped by the bits of knowledge acquired from the furtive get-together, they left determined to connect the holes that separated society, to repair the frayed strings of association that had unwound over the long run.

The principal challenge lay in defying the dug in power structures that propagated disparity. The corporate chief, when consumed by the tireless quest for riches, ended up attracted to the hero's vision of a general public where achievement was estimated not in money related gains but rather in the prosperity of all. Together, they looked to reclassify the motivation behind business, implanting it with a feeling of social obligation that rose above overall revenues.

All the while, the road craftsman loaned their imaginative gifts to the reason, utilizing their wall paintings to enhance the voices of the minimized. Walls that had once filled in as hindrances became materials for accounts of versatility and disobedience. The city, when isolated by financial inconsistencies, started to observe the development of a common story — a shared mindset that perceived the reliance of its occupants.

In any case, the excursion towards harmony was not without its difficulties. The hero confronted obstruction from the people who gripped to the recognizable solaces of the state of affairs, who dreaded the disintegration of limits that had long characterized their personalities. The powers of entropy tried to sabotage the hero's endeavors, planting dissension and doubt afterward.

Courageous, the hero dove further into the maze of cultural intricacies, searching out the main drivers of strife. All the while, they experienced people who had become lost despite any effort to the contrary of the framework, their accounts concealed underneath the facade of progress. The hero tuned in, their sympathy rising above the boundaries of bias and judgment.

In the neglected corners of the city, where the shadows of disregard posed a potential threat, the hero started grassroots developments that engaged the disappointed. Training turned into an encouraging sign, offering a pathway out of patterns of destitution and hopelessness. Local area gardens jumped up in deserted parts, cultivating a feeling of shared liability regarding the climate. The hero's activities undulated through the city, making pockets of change that tested the common account of division.

However, even as the hero created change on a nearby scale, the more extensive powers of worldwide unevenness cast an approaching shadow. The interconnectedness of the world, once praised for its true capacity for solidarity, presently appeared as a mind boggling snare of international strains, ecological debasement, and financial incongruities.

The hero, having seen the grandiose dance that bound all of presence, perceived the requirement for a comprehensive way to deal with address the difficulties that rose above city limits.

In quest for this more extensive vision, the hero turned into a worldwide messenger of balance. They navigated landmasses, drawing in with pioneers and visionaries from different societies, winding around an embroidery of coordinated effort that tried to blend the harsh notes of worldwide relations. The hero's revelation, once bound to the city's roads, presently resounded on a planetary scale.

Ecological stewardship turned into a point of convergence of the hero's worldwide drives. Perceiving the delicacy of the planet's biological systems, they supported for feasible practices that fit with the regular world. The once troublesome issue of environmental change turned into a mobilizing point for solidarity, as countries put away contrasts to address the common test of saving the planet for people in the future.

All the while, the hero attempted to destroy the obstructions that partitioned countries, encouraging a feeling of participation that rose above political philosophies. Tact, once portrayed by fights for control and lose situations, went through a change in outlook. The hero's vision of interconnectedness turned into a core value, cultivating coalitions that focused on the prosperity of mankind over tight public interests.

However, as the hero explored the complicated scene of worldwide issues, they experienced obstruction from settled in frameworks that gripped to obsolete ideal models. The powers of inactivity, energized by dread and personal responsibility, looked to subvert the hero's endeavors every step of the way. The battle for balance played out not just in the roads and hallways of force yet additionally inside the hearts and psyches of people whose perspectives were tested by the possibility of progress.

Unflinching, the hero drew strength from the shared mindset that had risen up out of the city roads. The interconnected organization of masterminds, scholars, and visionaries kept on developing, rising above geological limits to frame a worldwide local area of similar people. This shared perspective, powered by the hero's revelation, turned into an amazing powerhouse — an encouraging sign that enlightened the way towards an amicable world.

The excursion of Harmony Revelation, once bound to the singular's mission for understanding, had changed into a development that traversed the globe. The hero's revelation, refined into a straightforward yet significant truth — the interconnectedness of all things — had turned into an energizing weep for change. It reverberated through the corridors of force, resounded in the roads, and reverberated in the hearts of the people who longed for a reality where equilibrium and concordance won.

As the hero kept on exploring the intricacies of the worldwide scene, they ended up confronting a definitive trial of their convictions. An emergency of remarkable size unfurled — an intermingling of ecological disasters, international pressures, and social commotion that took steps to steer the results towards irreversible mayhem. The world remained at the cliff, wavering near the precarious edge of a calamitous disentangling.

Despite this existential danger, the hero's revelation expected another importance. The interconnectedness, everything being equal, when a core value for individual and cultural change, presently held the way to exploring the fierce waters of worldwide emergency. The hero, drawing upon the illustrations gained from the city roads and the hallways of force, set out on a mission to join mankind notwithstanding misfortune.

The hero's process took them to the bleeding edges of catastrophe, where the impacts of ecological debasement were revealed. Amidst crushed scenes and uprooted networks, the hero saw the versatility of the human soul. It was here, in the midst of the destruction of a world on the edge, that the genuine capability of interconnectedness uncovered itself.

Networks, once separated by lines and contrasts, met up in the soul of common guide. Countries put away verifiable complaints, perceiving the common weakness that rose above international competitions. The hero, when a solitary vagabond on the city roads, presently remained in charge of a worldwide development — an alliance of countries and people joined by a typical reason.

As the hero explored the difficulties of the worldwide emergency, they experienced reverberations of their past battles. The powers of entropy, tenacious in their quest for turmoil, looked to take advantage of the weaknesses uncovered by the joining emergencies. Political radicals, driven by a craving for power and control, looked to plant friction and division. The hero, mindful of the delicacy of harmony, stood up to these provokes with an unfaltering assurance to safeguard the interconnected web that kept the world intact.

In the cauldron of emergency, the hero's revelation went through a transformation. As of now not bound to the domain of individual comprehension or cultural change, it turned into an encouraging sign that enlightened the way towards a strong and amicable world. The interconnectedness, everything being equal, when a theoretical idea, presently appeared as an unmistakable power that bound humankind together notwithstanding misfortune.

The hero's process finished in a snapshot of aggregate arousing — a worldwide revelation that resonated across mainlands. In the repercussions of the emergency, as networks remade and countries produced new unions, the world arose changed. The old ideal models that had propagated division and difficulty disintegrated, clearing a path for another period characterized by participation and mutual perspective.

Harmony, when a delicate state compromised by the powers of unevenness, presently turned into a core value for the world's future. The hero, whose excursion had started on the city roads, presently remained as an image of the groundbreaking force of a singular's revelation. Their story, woven into the texture of a world renewed, turned into a demonstration of the potential for positive change that lived inside the human soul.

As the sun set not too far off, projecting a warm shine over the revived city and the restored planet, the hero pondered the excursion that had prompted this second. The revelation, when a flash inside their own cognizance, had touched off a worldwide fire of understanding and solidarity. The balance that had been so dubiously kept up with in the past currently stood braced by the aggregate will of mankind.

Thus, in the peaceful snapshots of reflection, as the hero looked at the changed world, a significant feeling of satisfaction washed over them. The balance revelation, brought into the world from the mind boggling dance of the universe and supported through the preliminaries of individual and worldwide change, had turned into a directing light for a world that had picked congruity over dissension, solidarity over division.

As the stars arose in the night sky, giving occasion to feel qualms about their shining light the world underneath, the hero realize that the excursion was not genuinely finished. The dance of balance, steadily advancing and dynamic, would proceed.

However, equipped with the insight acquired from their revelation, the hero embraced the continuous hit the dance floor with a feeling of direction and assurance.

For in the core of the balance revelation, they had found the interconnectedness of everything as well as the endless potential for humankind to shape its fate. Thus, with a recharged feeling of trust and a promise to the rules that had directed their excursion, the hero ventured forward into the strange domain representing things to come, prepared to confront anything that difficulties lay ahead.

8.1 After the confrontation, the group reflects on their journey and the lessons learned.

In the repercussions of the serious showdown that had tried the constraints of their purpose, the gathering wound up in an intelligent second. The air hung weighty with the reverberations of the new battle, and as the residue settled, they accumulated to mull over the way that had carried them to this urgent point. The excursion, an embroidery woven with strings of challenge, disclosure, and change, unfurled in their shared mindset.

Their appearance started with the beginning of their common mission, an apparently normal undertaking that had developed into a remarkable odyssey. Every individual from the gathering conveyed the heaviness of individual narratives and remarkable points of view, combining at a junction that requested solidarity notwithstanding misfortune.

The illustrations advanced en route turned into the establishments whereupon they fabricated their mutual perspective as well as the bonds that currently bound them as a tough group.

The gathering's process had started with a source of inspiration, a request that enticed them to wander past the bounds of the recognizable. As different people, they explored the unfamiliar domains of the obscure, drawn together by a mutual perspective that rose above the limits of their singular lives. The preliminaries that looked for them were both outer and interior, testing their actual perseverance as well as the profundities of their personality.

In the beginning phases, clashes emerged inside the gathering as different belief systems conflicted like structural plates underneath the surface. It was a cauldron of contrasts, a trial of their ability to accommodate different convictions for a shared objective. The excursion requested a recalibration of individual viewpoints, an affirmation that the strength of their solidarity would be manufactured in the flames of understanding.

As they navigated scenes set apart by vulnerability and uncertainty, the gathering experienced tutors and difficulties the same. Savvy guides showed up, granting immortal insights that rose above the prompt targets of their journey. These tutors became signals of intelligence, enlightening the way ahead and furnishing the gathering with bits of knowledge that reached out past the limits of their unique mission.

However, the excursion was not a direct movement toward an effectively characterized objective. It wandered through valleys of uncertainty and climbed pinnacles of disclosure, reflecting the undulating rhythms of the human experience. The gathering confronted snapshots of depression, wrestling with the monstrosity of the difficulties before them. It was at these times that the seeds of flexibility were planted, and the gathering found the strength that rose up out of their aggregate assurance to drive forward.

In their appearance, the gathering recognized the essential job of misfortune in molding their personality. The preliminaries they confronted were not inconsistent hindrances yet rather valuable open doors for development and self-disclosure. Every misfortune, every snapshot of vulnerability, turned into a cauldron where their courage was tried and refined. It was through difficulty that the gathering uncovered repositories of inward strength, producing a rugged bond that endured the tensions of outer difficulties.

The outside challenges, however considerable, were nevertheless impressions of the unseen struggles that every part wrestled with exclusively. The excursion, it became obvious, was not just an investigation of their general surroundings yet additionally a significant reflection into the openings of their own spirits. As they stood up to outer foes, they at the same time went up against the shadows inside — the questions, fears, and instabilities that took steps to sabotage their aggregate reason.

In their appearance, the gathering wondered about the extraordinary force of shared weakness. As they opened dependent upon each other, uncovering the profundities of their battles and desires, an embroidery of shared mankind arose. The walls that had once isolated them disintegrated, giving way to a significant feeling of interconnectedness. It was an acknowledgment that, regardless of their disparities, they were limited by an ongoing idea — a common humankind that rose above the names and classifications that frequently separated them.

The excursion likewise drove the gathering to scrutinize the actual idea of their journey. What, they considered, was the genuine pith of their central goal? Was it only the fulfillment of an unmistakable objective, or did the importance lie in the elusive changes that happened en route? The gathering wrestled with the acknowledgment that the actual excursion was the objective — that the examples took in, the associations produced, and the internal scenes investigated were the genuine fortunes they had uncovered.

In the pot of self-disclosure, the gathering defied their assumptions and predispositions. They scrutinized the presumptions that had molded their perspectives and stood up to the awkward insights that lay underneath the surface. It was a course of forgetting — a shedding of old ideal models that cleared a path for a more far reaching comprehension of themselves and the world they occupied.

The gathering's appearance stretched out to the essential snapshots of conflict that had carried them to this crossroads. The foes they confronted were not simple impediments but rather impressions of the difficulties that snuck inside the aggregate mind of humankind. The gathering perceived that the genuine fight was not against outer powers alone but rather against the shadows of obliviousness, lack of care, and division that cast a long shadow over the world.

The showdown had requested mental fortitude and versatility, characteristics that the gathering had developed through the preliminaries of their excursion. It was a fight not just for the safeguarding of their nearby objectives yet for the actual soul of the world they looked to change. In confronting outer enemies, they had likewise stood up to the murkiness inside, perceiving the cooperative connection between the individual and the system.

As the gathering mulled over the illustrations learned, they recognized the certainty of progress. The excursion had been a course of consistent development, a unique hit the dance floor with the steadily moving flows of presence. The gathering wondered about their own ability for transformation, understanding that the examples learned had prepared them for the difficulties of the past as well as for the vulnerabilities representing things to come.

In the calm snapshots of reflection, the gathering recognized the meaning of aggregate reason. Their excursion, once set out upon as people, had combine into a common account — a story that rose above the limits of their singular lives.

That's what the acknowledgment unfolded, as they continued looking for change, they had become draftsmen of a bigger story — a story that reached out past the bounds of their nearby objectives and enveloped the more extensive embroidery of human experience.

The gathering's appearance likewise addressed the idea of administration that had arisen naturally inside their middle. The conventional thoughts of power had given way to a more liquid and comprehensive type of initiative — one that perceived the qualities of every person and outfit the aggregate insight of the gathering. It was an authority brought about for a specific need, fashioned in the pot of shared difficulties, and grounded in the standards of common regard and joint effort.

In their thought of the excursion, the gathering found comfort in the information that they were in good company as they continued looking for change. The world, they understood, was populated by endless people and gatherings took part in their own odysseys of self-revelation and cultural change. The gathering's excursion, how-ever special, turned into a microcosm of a bigger development — a development that tried to introduce another period of cognizance and interconnectedness.

The gathering's appearance developed as they thought about the effect of their ex-cursion on their general surroundings. The changes that had occurred inside every part undulated outward, impacting the networks they contacted and motivating others to leave on their own journeys for positive change. The gathering became impetuses for a more extensive development — a development that rose above geographic limits and social partitions.

The illustrations gained from their process became core values for the gathering's continuous endeavors. They perceived the requirement for consistent reflection, recognizing that the advancement of oneself and the aggregate was a continuous cycle. The gathering focused on excess cautious against the powers of carelessness and stagnation, understanding that the excursion towards change was an unending pattern of development and recharging.

As the gathering embraced the certainty of progress, they likewise recognized the force of trust. In the haziest snapshots of their excursion, trust had been a signal that directed them through the shadows. It was an acknowledgment that, regardless of the horde challenges that lay ahead, the aggregate soul of humankind had the ability to conquer difficulty and shape a more promising time to come.

The gathering's appearance closed with a feeling of appreciation — for the difficulties that had tried their strength, the tutors who had enlightened their way, and the aggregate excursion that had bound them together. The excursion, they got it, was an honor — a gift that offered not just the chance for individual and aggregate change yet in addition the opportunity to add to the more extensive embroidery of human development.

In the peaceful repercussions of reflection, the gathering stood joined together, braced by the illustrations learned and the bonds produced along their groundbreaking process. The reverberations of their aggregate revelation resounded through the air, conveying with them the commitment of a reality where people and networks, limited by a mutual perspective, could explore the intricacies of presence with versatility, sympathy, and a faithful obligation to positive change.

8.2 The protagonist comes to terms with their identity and the responsibility that comes with their power.

In the cauldron of self-revelation, the hero ended up at the convergence of character and power. The excursion that had unfurled before them, set apart by preliminaries and disclosures, had prompted a significant retribution with the pith of what their identity was and the heaviness of the capacities they had. As they remained on the edge of understanding, the hero wrestled with the double features of personality — the interior scene of mindfulness and the outer indication of force.

The excursion towards self-revelation had not been a straight movement but rather a confounded investigation of the intricacies that characterized the hero's personality. They dove into the openings of their own awareness, defying the layers of molding, cultural assumptions, and individual narratives that had formed their identity. It was an excursion that requested boldness — the fortitude to face not just the parts of themselves that they valued yet additionally the shadows that prowled toward the edges of their mind.

As the hero unwound the strings of their character, they experienced snapshots of disclosure and distress. The names that had once given a similarity to solidness currently felt choking, and the assumptions forced by society seemed like sick fitting pieces of clothing. The excursion requested a shedding of these outer develops, a stripping away of layers to uncover the crude realness underneath. It was a course of becoming untethered from the deceptions that had characterized their personality and a readiness to embrace the weakness that lay underneath the surface.

In the peaceful snapshots of contemplation, the hero faced questions that rose above the trivialities of cultural names. Who were they past the jobs allocated to them by others? What values and convictions resounded with the center of their being? The responses, tricky and consistently moving, became guideposts in the strange region of self-disclosure. The hero perceived that personality was not a decent point but rather a dynamic, developing substance that unfurled as time passes.

The excursion into self-revelation, in any case, was not a single undertaking. En route, the hero experienced mirrors as connections — associations that held up their very own impression pith. These connections, whether familial, dispassionate, or heartfelt, became impetuses for self-understanding.

The hero explored the intricacies of affection and misfortune, happiness and distress, perceiving that every communication left a permanent engraving on the material of their character.

It was through the cauldron of connections that the hero additionally defied the intrinsic obligation that accompanied their power. The capacities they had, whether physical, scholarly, or close to home, helped results that undulated through the embroidered artwork of their associations. The hero wrestled with the effect of their words and activities, understanding that the power they used could shape fates and impact the existences of people around them.

The outside indication of force, in any case, was not without its difficulties. The hero confronted the examination of a world that frequently misconstrued or looked to take advantage of their capacities. The obligation they bore weighed weighty on their shoulders — an obligation to involve their power for everyone's benefit, to elevate as opposed to persecute, and to explore the sensitive harmony among independence and interconnectedness.

As the hero explored the outside scene of force, they additionally went up against the inside components of obligation. Their decisions, informed by their advancing comprehension of self, reverberated through the passageways of their own soul. The hero wrestled with the results of their choices, perceiving that the ethical compass directing their activities held significant ramifications for the direction of their own excursion.

In the cauldron of character and power, the hero confronted snapshots of moral vagueness. The shades of dark that penetrated their decisions turned into a proving ground for the rules that secured their personality. They explored the landscape of moral issues, perceiving that the obligation that went with their power requested an increased consciousness of the potential for both positive and adverse consequence.

The excursion towards self-revelation and the acknowledgment of obligation unfurled against the background of cultural assumptions and fundamental elements. The hero, as a carrier of force, existed inside a bigger setting — a setting formed by social standards, verifiable heritages, and power structures that reached out a long ways past the limits of individual cognizance. The battle for credibility became laced with a more extensive journey for cultural change.

In their appearance, the hero wrestled with the multifacetedness of their character — the manners by which parts of orientation, race, and financial foundation met with and impacted their experience of force and obligation. The excursion drove them to defy their own inclinations and honor, testing assumptions and encouraging a more profound comprehension of the interconnected trap of social elements.

As the hero dove into the intricacies of cultural designs, they perceived the basic for support and allyship. The obligation that accompanied their power stretched out past the individual domain into the aggregate — an acknowledgment that the mission for equity and value was an inherent piece of their excursion. The hero turned into a voice for the minimized, a boss for those whose characters met with frameworks of persecution.

The excursion additionally provoked the hero to go up against the shadows of their own honor. They explored the distress that emerged from recognizing the benefits stood to them by cultural designs, perceiving that destroying these designs expected a readiness to surrender unmerited benefits. The hero wrestled with the strain between recognizing their own battles and perceiving the more extensive fundamental shameful acts that propagated patterns of imbalance.

In their consideration of personality and power, the hero embraced the idea that genuine strength lay not in mastery but rather in weakness and sympathy. The obligation they bore was not a weight to be carried alone however a common obligation to making a reality where power was employed with sympathy and equity. The hero comprehended that the excursion towards cultural change required a continuous exchange with others — an encouragement to all in all destroy severe designs and co-make a more fair future.

As the hero grappled with the complexities of their personality and the obligations of their power, they perceived the advantageous connection among self and society. The excursion, when an individual odyssey, had changed into an aggregate undertaking — a development towards realness and equity that resounded on both individual and cultural levels. The hero turned into an influencer, for themselves as well as for the more extensive woven artwork of humankind.

In the pot of self-disclosure and cultural change, the hero likewise wrestled with the fleetingness of personality. The acknowledgment unfolded that the excursion was a ceaseless pattern of development and development — a continuous course of shedding old skins and embracing new features of self. The hero explored the recurring pattern of character, understanding that the journey for credibility was a unique hit the dance floor with the steadily changing flows of presence.

The outer indications of force, as well, went through a transformation. The's comprehension hero might interpret liability extended past the quick circles of impact to incorporate a worldwide point of view. The difficulties of the world, whether natural, political, or compassionate, became fundamental parts of their excursion. The hero perceived that the obligation they bore stretched out to the more extensive embroidery of presence — an acknowledgment that filled their obligation to impact positive change on a worldwide scale.

In the calm snapshots of consideration, the hero recognized the meaning of heritage. Their decisions, the effect they had on the world, became strings woven into the texture of time.

The obligation they conveyed was not exclusively to the present yet additionally to people in the future — an acknowledgment that the excursion towards self-disclosure and cultural change was an immortal undertaking with extensive ramifications.

As the hero embraced the intricacy of their personality and the gravity of their power, they remained at the intersection of past, present, and future. The excursion, a mosaic of difficulties and revelations, had prompted a significant combination of self and cultural obligation. The hero, presently a steward of legitimacy and equity, ventured forward into the unfurling parts of their story, prepared to explore the steadily developing scenes of personality, power, and the aggregate journey for a more fair world.

8.3 Tie up loose ends and prepare for the resolution of the overarching plot.

In the embroidery of the overall plot, the story strings had woven a perplexing and complicated design, a mosaic of characters, clashes, and unsettled questions. As the story moved toward its conclusion, the opportunity had arrived to take care of potential issues and set up for the goal that anticipated not too far off. The characters, each bearing the heaviness of their singular circular segments, joined in a combination that held the commitment of replies, conclusion, and the satisfaction of fates.

The primary remaining detail to be tended to was the confounding figure that had waited in the shadows, coordinating occasions with a puppeteer's artfulness. The strange manikin ace, whose inspirations had stayed tricky, rose up out of the account shadows. Their personality, once clouded by layers of interest, was uncovered, uncovering an intricate embroidery of inspirations, wants, and a past intertwined with the predeterminations of the fundamental characters.

As the hero stood up to the manikin ace, an exchange unfurled — a talk that stripped back the layers of mystery and uncovered the complexities of the great plan that had molded the story. Inspirations, when covered in equivocalness, were uncovered, and the manikin expert's part in the characters' lives became more clear. It was a snapshot of disclosure, a critical point that enlightened the underpinnings of the overall plot.

The goal of the manikin expert's storyline was not without its intricacies. Their activities, however determined by a one of a kind arrangement of inspirations, had gotten rolling a chain of occasions that resonated through the existences of the principal characters. The showdown was not just a conflict of wills but rather an investigation of the interconnectedness that bound the characters to a common predetermination. The manikin ace, when a mystery, turned into an impression of the heroes' own excursions — a mirror wherein they saw the expected results of their own decisions.

As the account dug into the complexities of character connections, another remaining detail arose — the unsettled struggles and implicit strains that had stewed underneath the surface. The connections that had endured storms and confronted preliminaries were given space for goal. Characters went up against the apparitions of their past, tending to complaints, and looking for conclusion in snapshots of profound therapy.

The heartfelt traps that had added layers of intricacy to the story were likewise brought to a purpose in retribution. The unsettled sentiments, the waiting questions, and the implicit insights were revealed. The characters explored the complexities of affection, double-crossing, and absolution, wrestling with the flaws that made their connections both delicate and versatile. It was an embroidery of feelings — a summit of the heartfelt subplots that had woven through the general plot.

In the peaceful snapshots of reflection, characters additionally defied the ethical issues that had accentuated their excursions. The decisions they had made, the trade offs struck, and the outcomes of their activities were examined from the perspective of a shared mindset. The last details of moral uncertainty were not attached with slick retires from unconditional, welcoming the peruser to examine the shades of dim that hued the characters' choices.

The all-encompassing plot likewise required an assessment of the outside clashes that had pushed the account forward. The unsettled pressures between groups, the approaching dangers that cast shadows over the characters' lives, and the secrets that still couldn't seem to be unwound requested goal. The hero, equipped with freshly discovered understanding and a feeling of direction, left on a last journey to destroy the powers that looked to upset the fragile harmony of the story world.

The goal of outside clashes was not accomplished through beast force alone yet through the hero's capacity to explore the intricacies of discretion, coalition building, and a nuanced comprehension of the foes' inspirations. It was a demonstration of the development of the hero's personality — an acknowledgment that genuine power lay in actual ability as well as in the insight to recognize when to use power and when to expand a hand in compromise.

As the story strings merged towards the goal, the general plot requested a retribution with the topics that had moored the story. The investigation of personality, power, obligation, and the complexities of the human experience arrived at its apex. The characters, each a vessel for these overall topics, wrestled with the illustrations took in, the changes gone through, and the insight acquired through the cauldron of their separate processes.

The peak of the general plot unfurled in a crescendo of pressure and feeling. The hero, encompassed by partners manufactured through preliminaries, confronted the last enemy — a representative exemplification of the difficulties that had characterized their excursion.

The showdown was not only a conflict of physical could however a clash of belief systems, a battle for the spirit of the story world.

In the climactic minutes, the last details of the overall plot were drawn together. The disclosures, clashes, and goals unfurled in an embroidery of story conclusion. The manikin expert's inspirations were uncovered, the characters faced the phantoms of their past, and the outside clashes that had molded the story scene were either settled or set on a way towards goal. It was a snapshot of therapy, a perfection of the story components that had been carefully created all through the story.

However, even in the goal, the general plot left space for equivocalness — an acknowledgment that not all questions required slick responses and that the human experience was much of the time described by an absence of outright sureness. The last details that stayed loosened turned into a demonstration of the intricacy of the story world, welcoming the peruser to participate in their own understandings and reflections.

In the consequence of the goal, the story entered the end result — a period of reflection and repercussions that permitted the characters to deal with the situation that had unfolded. The repercussions was not a quick decision but rather a progressive loosening up of the story pressure, a space for characters to find a sense of peace with the ramifications of the goal and the changes that had happened inside themselves.

The outcome additionally gave a chance to investigate the results of the general plot on the account world. The cultural movements, the outcome of struggles, and the waves of progress that exuded from the characters' decisions were inspected. The story world, when set apart by pressure and vulnerability, entered another time formed by the goals of the general plot.

As the characters explored the end result, they likewise set out on private epilogs — individual reflections on the excursion they had attempted and the ways that lay ahead. The last details of their singular bends tracked down goal, and the characters faced the vulnerability representing things to come with a freshly discovered feeling of organization. The conclusion turned into a material for individual person development, a space for them to rise up out of the shadows of the overall plot and step into the radiance of their own predeterminations.

The goal of the all-encompassing plot, while giving conclusion to the quick story clashes, additionally indicated the coherence of the characters' excursions. The conclusion of the goal was tempered by the acknowledgment that life, both in the account world and then some, was a continuous course of development and advancement.

The overall plot, however closed, turned into a basic section in the characters' bigger stories — a part that would illuminate their decisions, shape their predeterminations, and wait in the reverberations of their shared mindset.

In the last snapshots of the account, the remaining details of the all-encompassing plot were not flawlessly attached with bows however left to wait in the peruser's creative mind. The unsettled inquiries, the unassuming prospects, and the waiting vagueness turned into a greeting for reflection — an affirmation that the genuine excellence of narrating frequently lay not in the clean goals but rather in the spaces hidden therein, where perusers could draw in their own imagination and translation.

As the drape fell on the general plot, the story world, once laden with pressure and vulnerability, entered another part. The characters, having defied their devils and explored the intricacies of their individual processes, remained at the limit of a future formed by the goals of the past. The last details, however not all flawlessly tied, turned into an embroidery of story wealth — a demonstration of the intricacy, profundity, and getting through secrets of the general plot.

Chapter 9

The Eternal Flame

In the core of the supernatural domain of Eldoria, where the texture of the truth is woven with strings of enchantment and secret, there exists a peculiarity referred to every one of its occupants as the Everlasting Fire. It is a divine puzzle, a brilliant signal that has endured over the extreme long haul, glimmering with a supernatural splendor that rises above the comprehension of mortal personalities.

Legends and fantasies entwine with the actual pith of the Everlasting Fire, winding around an embroidery of stories that have been gone down through ages. Eldorians, the occupants of this ethereal land, view the Fire as an image of congruity, an indication of the everlasting pattern of life and sorcery.

The beginnings of the Everlasting Fire are hidden in the fogs of time, lost in the records of Eldoria's antiquated history. Some accept that it was a gift gave to the domain by the divine beings themselves, a heavenly flash that lighted the actual soul of the land. Others guarantee that it arose out of the union of strong ley lines, drawing upon the crude energy that courses through the veins of the universe.

No matter what its beginnings, the Timeless Fire has turned into a point of convergence for the occupants of Eldoria. It is said that the Fire has the ability to uncover stowed away insights, to enlighten the most obscure corners of the heart, and to direct the individuals who look for its insight. Pioneers from all edges of the domain set out on misleading excursions to observe the Fire's entrancing dance and to loll in its brilliant gleam.

The journey to the Everlasting Fire is no simple accomplishment. It requires crossing tricky scenes, overcoming charmed backwoods, and exploring supernatural mazes that resist the actual laws of existence. Just not set in stone and unadulterated of heart can expect to arrive at the sacrosanct grounds where the Fire dwells.

As one sets out on the challenging excursion, stories of the Fire's wonders and preliminaries reverberation through the ears of the voyagers like tormenting tunes. It is said that the people who approach with truthfulness and modesty are conceded dreams of the past, brief looks into the future, and bits of knowledge into the mind boggling embroidery of predetermination that ties each living being in Eldoria.

The excursion to the Timeless Fire isn't just an actual odyssey yet an other-worldly one too. En route, travelers experience legendary animals, big-hearted spirits, and malicious elements that test the guts of their purpose. The Fire, it is accepted, passes judgment on the value of the people who look for its presence, and just the unadulterated of heart are permitted to remain before its brilliant greatness.

Legends recount a picked rare sorts of people who have communed with the Ever-lasting Fire and arisen as vessels of its power. These people, known as the Flamebearers, are gave with a small portion of the Fire's pith, giving them uncommon capacities and a profound association with the mysterious flows that course through Eldoria.

The Flamebearers, respected as living channels of the Timeless Fire, are depended with the consecrated obligation of safeguarding the equilibrium of sorcery in the domain. They are the gatekeepers of antiquated information, managers of enchanted curios, and safeguards against the infringing powers that try to disentangle the sensi-tive strings of Eldoria's presence.

In any case, to whom much is given, much will be expected, and the Flamebearers are not absolved from the difficulties that go with their heavenly gifts. The enchanted that courses through their veins is a situation with two sides, fit for both creation and obliteration. As they explore the complex dance among light and shadow, the Flamebearers should wrestle with the heaviness of predetermination and the penances expected to maintain the holiness of Eldoria.

Over the ages, the Flamebearers have made a permanent imprint on the woven artwork of Eldoria's set of experiences. Their stories are sung in numbers, carved into antiquated books, and murmured by the breeze through the ages.

Some Flamebearers became amazing legends, driving armed forces to triumph against dim powers that took steps to overwhelm the domain. Others, consumed by the intense wizardry that flowed through their veins, capitulated to the charm of taboo information and went wrong, abandoning a tradition of misfortune and wariness.

One such story is that of Aeliana, a Flamebearer of unrivaled beauty and shrewd-ness. Her excursion to the Timeless Fire was loaded with risk, as she explored through the shadowy domains and confronted preliminaries that tried the restrictions of her solidarity and strength. Aeliana's heart ignited with an intense longing to safeguard Eldoria from the infringing haziness, and her association with the Fire extended with each step of her odyssey.

Aeliana's predetermination became entwined with an old prediction that predicted an approaching disturbance taking steps to overwhelm Eldoria in timeless evening. The Fire murmured mysteries to her in the hallowed tongue of wizardry, uncover-ing the way she should track to deflect the approaching destruction. With faithful assurance, she accumulated partners from divergent corners of the domain — heroes, researchers, and otherworldly creatures joined by a typical reason.

The excursion to save Eldoria unfurled as a fabulous epic, with Aeliana and her partners confronting legendary monsters, opening the insider facts of failed to remember domains, and unwinding the secrets that bound the domain together. The Flamebearers' novel capacities became instrumental in their journey, as they bridled the force of the Timeless Fire to recuperate wounds, control components, and even look into what's in store.

However, as the Flamebearers moved nearer to their last a conflict with the powers of murkiness, the limits among partner and enemy obscured. Double-crossings and penances became inescapable, as the line among light and shadow turned out to be progressively dubious. Aeliana, troubled by the heaviness of initiative and her information that the decisions would reverberate until the end of time, wrestled with the ethical predicaments that went with employing the force of the Everlasting Fire.

The climactic fight against the powers of haziness unfurled underneath the approaching shadow of the divine sky. The actual texture of Eldoria shuddered as the restricting energies conflicted, making an ensemble of wizardry that resounded across the domains. Aeliana, filled by the quintessence of the Timeless Fire, remained at the front of the contention, confronting the encapsulation of the approaching disturbance.

In the pot of the last fight, the Flamebearers released the full degree of their powers, winding around spells that rose above the limits of mortal comprehension. The air popped with mystical energy as the destiny of Eldoria remained in a critical state. Aeliana, her eyes burning with the splendor of the Everlasting Fire, defied the wellspring of the infringing haziness with a boldness manufactured in the pot of her excursion.

As the powers crashed in a disastrous crescendo, Aeliana's association with the Fire arrived at its peak. The Flamebearer turned into a vessel of unadulterated enchantment, rising above the restrictions of her human structure. In that extraordinary second, she communed with the actual heart of Eldoria, the throbbing center of wizardry that resounded with the Everlasting Fire.

In a flood of brilliant energy, Aeliana outfit the force of the Timeless Fire to expel the obscurity that took steps to consume Eldoria. The disastrous powers withdrew, crushed by the unfaltering light of the Fire. The domains shuddered as the equilibrium was reestablished, and the natives of Eldoria viewed Aeliana with love, for she had turned into a living encapsulation of their salvation.

However, the triumph included some significant downfalls. Aeliana, her human structure incapable to contain the staggering enchantment flowing through her, rose to a higher plane of presence. Her quintessence converged with the Everlasting Fire, turning into a piece of the heavenly dance that enlightened the embroidery of Eldoria. She turned into a directing light, a signal for people in the future, and a demonstration of the getting through force of penance and magnanimity.

The story of Aeliana and the Flamebearers turned into a foundation of Eldoria's legend, a story went down through ages as a sign of the delicate harmony among light and shadow. The Everlasting Fire kept on glinting, undiminished by the progression of time, its brilliance a demonstration of the unstoppable soul of Eldoria and the timeless pattern of sorcery that bound the domain together.

As the ages unfurled, new Flamebearers arose, each with their own fates and difficulties. The journey to the Everlasting Fire persevered, as searchers of truth and shrewdness set out on ventures that reflected the preliminaries of the people who preceded them. The Fire, everlasting and unfaltering, stayed an image of expectation and motivation for all who looked at its entrancing dance.

Thus, in the core of Eldoria, where enchantment and secret joined, the Everlasting Fire kept on consuming, an immortal demonstration of the persevering through soul of a domain limited by the strings of predetermination. However long there were Flamebearers to maintain the heritage, and travelers able to set out on the hallowed excursion, the Fire would persevere, projecting its brilliant shine across the domains and enlightening the way to a future saturated with sorcery and marvel.

9.1 Conclude the story with a resolution to the larger conflict and the fate of the world.

In the consequence of Aeliana's otherworldly penance, Eldoria lounged in the newly discovered light that exuded from the Everlasting Fire. The waiting reverberations of the calamitous fight continuously blurred, and the domain started to recuperate from the scars left by the infringing obscurity. The occupants of Eldoria, from the captivated timberlands to the supernatural urban communities, felt the unmistakable change in the mystical flows that wove through the texture of their reality.

The Flamebearers, who had battled close by Aeliana in the climactic showdown, ended up different by the experience. The leftover substance of the Timeless Fire waited inside them, a demonstration of their association with the divine powers that had directed them on their hazardous excursion. Some battled to outfit the freshly discovered powers, wrestling with the heaviness of obligation that accompanied being channels of such intense sorcery. Others embraced their jobs with a feeling of obligation, promising to safeguard Eldoria and maintain the tradition of the individuals who had preceded them.

As the domains recuperated, the Flamebearers accumulated to think about the illustrations picked up during their odyssey. The Timeless Fire, presently like never before, filled in as a signal of direction, offering experiences into the sensitive equilibrium that kept Eldoria intact. They understood that the battle among light and shadow was not a limited fight but rather an unending dance, an ebb, and stream that expected everlasting cautiousness.

Directly following the upheaval, a board of Flamebearers arose to manage the conservation of Eldoria's wizardry and the defending of its magical legacy. They met in the consecrated forests where the Timeless Fire consumed most brilliant, trading shrewdness and producing a settlement to guarantee that the illustrations of the past were not neglected. Their solidarity turned into an image of the flexibility of Eldoria, a demonstration of the strength that could be drawn from variety and common perspective.

In the domains past the charmed forests, individuals of Eldoria saw the unmistakable impacts of the Fire's resurgence. Crops thrived, the timberlands overflowed with dynamic life, and the very air gleamed with a rejuvenated energy. The once-muffled melody of enchantment resounded through the land, making an agreeable orchestra that repeated the interconnectedness of every living thing.

However, even directly following triumph, murmurs of a waiting danger waited on the breeze. The powers that had looked to dive Eldoria into everlasting night were not completely vanquished. Shadows actually stuck to the edges of the domain, remainders of pernicious elements that waited in the secret corners of the real world.

The Flamebearers, sensitive to the unpretentious subtleties of the mysterious flows, detected the remainders of murkiness and perceived that their assignment was not even close to finished. Aeliana's penance had pushed back the infringing tide, yet the timeless dance among light and shadow requested interminable carefulness. Yet again the board assembled, thinking on the means expected to brace Eldoria against possible future dangers.

In their conversations, the Flamebearers wrestled with the sensitive harmony among request and tumult, perceiving that limits in either heading could steer the results towards lopsidedness. They looked to wind around an embroidery of concurrence, where wizardry streamed uninhibitedly however was tempered by the insight of the individuals who grasped its many-sided designs.

One Flamebearer, a sage named Elandra, proposed the making of mystical wards and charms that would go about as a defensive hindrance against vindictive powers. Drawing upon the aggregate information on the committee, she illustrated an arrangement to inject the actual embodiment of the Timeless Fire into the texture of Eldoria itself, making an agreeable reverberation that would repulse any dimness that thought for even a second to infringe.

The gathering, perceiving the insight in Elandra's proposition, set about carrying out the arrangement. The Flamebearers headed out to the furthest reaches of Eldoria, diverting their aggregate wizardry into the production of magical wards that spread over the broadness of the domain. Each ward was a demonstration of the Flamebearers' devotion, a sign of their obligation to shielding the sensitive equilibrium that kept Eldoria intact.

The interaction was not without challenges. The making of the wards expected the Flamebearers to take advantage of the actual embodiment of the Everlasting Fire, an undertaking that tried their dominance over the strong sorcery that moved through their veins. The domains trembled with the force of their endeavors, and the very air snapped with the reverberation of divine energies.

As the wards came to fruition, Eldoria itself appeared to answer. The land consumed the enchanted like a living life form, incorporating the quintessence of the Everlasting Fire into its actual center. The mysterious obstructions turned into a consistent piece of the domain, imperceptible to the unaided eye however reverberating with a strong charm that repulsed murkiness and vindictiveness.

With the wards set up, the Flamebearers stood joined in their determination to safeguard Eldoria. The board, drove by Elandra, proclaimed that the Flamebearers would proceed with their careful watch over the domains, guaranteeing that the sensitive equilibrium they had reestablished would persevere through the ages.

Yet again individuals of Eldoria, flourishing under the altruistic look of the Everlasting Fire, commended the arrival of success and sorcery to their lives. The story of Aeliana and the Flamebearers turned into a foundation of Eldoria's shared perspective, a story went down through ages to move trust and mental fortitude in the midst of difficulty.

The Flamebearers, presently adored as savvy stewards of sorcery, became tutors and guides for the people who tried to grasp the magical flows that molded their reality. The journey to the Everlasting Fire continued, not as a frantic mission to deflect upheaval, but rather as a representative excursion of self-revelation and illumination. Travelers from varying backgrounds set out on the odyssey, trying to associate with the everlasting dance of wizardry and to draw motivation from the Flamebearers' heritage.

As the patterns of time unfurled, Eldoria remained as a demonstration of the persevering through force of versatility and solidarity. The Flamebearers, limited by a mutual perspective, kept on passing down their insight through the ages. The wards, woven into the actual texture of the domain, remained as quiet sentinels, gatekeepers against the infringing shadows that unendingly tried to test the domain's determination.

The domains flourished, and the reverberations of the past turned into a directing power for people in the future. The Everlasting Fire, its brilliant shine undiminished, glimmered with an immortal brightness that rose above the limits of mortal comprehension. Eldoria, a domain molded by the fragile dance among light and shadow, remained as a signal of wizardry and miracle, a demonstration of the getting through soul that bound its natives together.

Thus, in the core of Eldoria, where the otherworldly energies united, the Everlasting Fire kept on consuming, projecting its brilliant sparkle across the domains. The story of Aeliana and the Flamebearers turned into a remarkable individual, a story carved into the actual soul of the land. However long there were Flamebearers to maintain the heritage, and explorers able to set out on the sacrosanct excursion, the Fire would persevere, enlightening the way to a future saturated with wizardry and marvel.

9.2 Reflect on the personal growth of the protagonist and the impact of their choices.

Aeliana's excursion, from a conventional occupant of Eldoria to a Flamebearer of extraordinary power, unfurled as a legendary story of self-improvement and penance. The preliminaries she confronted, the decisions, and the weights she bore all added to a significant change that undulated through the actual texture of her being.

At the start of her odyssey, Aeliana was nevertheless an unassuming soul, her reality entwined with the recurring pattern of regular day to day existence in Eldoria. The call of the Everlasting Fire, nonetheless, coaxed to something more profound inside her — an inert likely that rose above the limits of the unremarkable.

Her initial steps on the journey were provisional, driven by interest and a longing for a more prominent comprehension of the magical powers that molded her reality.

As Aeliana navigated the deceptive scenes and confronted the legendary animals that monitored the way to the Timeless Fire, she experienced difficulties that tried her actual ability as well as the profundities of her personality. The excursion turned into a cauldron, producing her versatility and supporting a feeling of assurance that would demonstrate fundamental in the preliminaries to come.

The experiences with generous spirits and malicious elements the same filled in as mirrors, mirroring the features of Aeliana's own spirit. The Fire, receptive to the immaculateness of her goals, offered looks into the openings of her heart, uncovering weaknesses, fears, and the undiscovered repositories of solidarity that lay lethargic inside. Every disclosure turned into a venturing stone on the way of self-revelation, directing Aeliana towards a comprehension of the significant association between her inward world and the enchanted that coursed through Eldoria.

The connections produced during her process assumed an essential part in Aeliana's self-improvement. Partners, attracted to her steady determination and the flash of the Fire inside, joined her mission. Together, they endured the hardships of vulnerability, confronted difficult chances, and framed bonds that rose above the limits of simple fellowship. These partners, each with their own assets and shortcomings, turned into an impression of the aggregate soul expected to stand up to the approaching haziness.

One such partner was Aric, a carefully prepared fighter with a spooky past. As their ways united, Aeliana tracked down in him a close companion — a spirit scarred by the shadows of bygone eras yet powered by a craving for recovery. Aric's process reflected Aeliana's in numerous ways, and their common encounters turned into a pot that tempered their fellowship into a bond strong even despite difficulty.

Through Aric, Aeliana took in the craft of strength and the force of pardoning. The Fire, ever careful, perceived the extraordinary effect of their brotherhood, mixing their partnership with a cooperative energy that rose above individual qualities. Aric's unflinching faithfulness and enduring help filled in as a wellspring of motivation for Aeliana, supporting her confidence in the potential for recovery and the strength that lay in solidarity.

However, as the odyssey unfurled, Aeliana likewise confronted snapshots of significant isolation. In the magical mazes and charmed backwoods, where the murmurs of the Fire were her main buddies, she dove profound into the openings of her own spirit. The isolation turned into an educator, constraining Aeliana to defy her most profound feelings of trepidation and uncertainties, to scrutinize the inspirations that drove her, and to track down comfort in the calm snapshots of thoughtfulness.

In those snapshots of isolation, the Flamebearer found the idle power that dwelled inside her — the ability to shape her fate and impact the actual texture of Eldoria. Her decisions, filled by the lucidity that isolation offered, undulated through the domains like a stone cast into a still lake. The Fire, a quiet observer to her choices, answered with a dance of ethereal light, repeating the significant effect of individual decisions on the fantastic embroidery of predetermination.

The decisions were not without outcome, and Aeliana wrestled with the heaviness of obligation that went with her newly discovered mindfulness. The power she employed, drawn from the pith of the Timeless Fire, was a situation with two sides equipped for both creation and obliteration. The Fire, timeless and untamed, requested a fragile dance — an agreeable harmony between employing sorcery for everyone's benefit and opposing the enticements that hid in the shadows.

As the odyssey arrived at its peak, Aeliana remained at the incline of predetermination, confronting a decision that would decide the destiny of Eldoria. The powers of murkiness, epitomized in an approaching disaster, took steps to dive the domain into everlasting evening. The Flamebearer, presently a vessel of divine energy, wrestled with the hugeness of the choice before her — a demonstration of generosity that would blend her embodiment with the Everlasting Fire to expel the infringing dimness.

The heaviness of the decision pushed ahead upon Aeliana's shoulders, and the Fire murmured insights that rose above the limits of mortal comprehension. In that critical second, the Flamebearer's self-improvement arrived at its peak, as she embraced the mantle of fate with a fortitude manufactured in the pot of her excursion. The penance, significant and caring, turned into a demonstration of the extraordinary force of decisions made in the cauldron of misfortune.

As Aeliana converged with the Timeless Fire, her actual structure rose above the restrictions of mortality. The brilliant energy, a mixture of her quintessence and the divine force of the Fire, turned into a signal that ousted the haziness and reestablished harmony to Eldoria. The penance reverberated through the domains, an ensemble of wizardry that resounded with the actual heart of the land.

Aeliana's preferred effect resonated a long ways past the domains she tried to safeguard. The Flamebearers, presently inheritors of her heritage, demonstrated the veracity of the groundbreaking force of penance and the everlasting dance among light and shadow. The supernatural wards, woven into the texture of Eldoria, turned into a demonstration of the persevering through soul that bound the domain together — a mind boggling embroidery of wizardry and miracle, molded by the decisions of the people who thought for even a second to remain against the infringing haziness.

Individuals of Eldoria, recipients of Aeliana's penance, felt the unmistakable impacts of her decisions. Thriving got back to the land, the mystical flows resounded with recently discovered imperativeness, and the very air sparkled with the quintessence of the

Timeless Fire. Aeliana's story turned into a remarkable person, a wellspring of motivation for people in the future who tried to figure out the groundbreaking force of decisions and the fragile dance between individual organization and the fabulous embroidery of fate.

Following Aeliana's penance, the Flamebearers assembled to consider the examples learned. The board, presently depended with the stewardship of Eldoria's sorcery, wrestled with the obligations that accompanied their newly discovered jobs. The Fire, everlasting and all-knowing, directed their thoughts, offering experiences into the fragile equilibrium expected to maintain the tradition of Aeliana and the Flamebearers who had preceded.

The effect of self-improvement, manufactured through the pot of decision and difficulty, turned into a directing power for the Flamebearers. The examples gained from Aeliana's process filled in as a compass, guiding them towards a future where the fragile dance among light and shadow was never-endingly maintained. The Fire, a quiet spectator and timeless friend, kept on igniting with undiminished splendor, projecting its brilliant sparkle across the domains.

The self-awareness of Aeliana, reflected in the decisions of the Flamebearers who followed, turned into a demonstration of the persevering through soul of Eldoria. The odyssey, with every one of its hardships, unfurled as an account of change — a story woven into the actual soul of the land. However long there were Flamebearers to maintain the heritage, and explorers ready to leave on the sacrosanct excursion, the Fire would persevere, enlightening the way to a future saturated with wizardry and marvel.

9.3 Leave room for potential sequels or spin-offs, hinting at new adventures in the world of temperature manipulation.

In the domain of Eldoria, where wizardry and secret joined in a perplexing dance, another part unfurled following Aeliana's penance. The iridescent sparkle of the Everlasting Fire, undiminished by time, washed the land in a brilliant hug. The otherworldly wards, woven into the actual texture of Eldoria, remained as quiet sentinels against the infringing obscurity. The Flamebearers, inheritors of Aeliana's heritage, gathered in the hallowed forests to ponder the examples learned and the obligations that lay ahead.

Among the Flamebearers, a figure arose whose predetermination was interwoven with an extraordinary and unfamiliar part of wizardry — temperature control. Lyra, a youthful proficient with an intrinsic capacity to control intensity and cold, turned into a guide of interest for her companions. Her process reflected Aeliana's, as she set out on a journey to the Timeless Fire, looking for direction on the neglected boondocks of temperature control.

As Lyra navigated the captivated scenes and confronted the preliminaries that looked for her, she found the undiscovered possibility of her capacities. The Fire, ever vigilant, murmured insider facts to her, noteworthy the sensitive dance among fire and ice that could shape the actual pith of Eldoria.

The Flamebearers, perceiving the meaning of Lyra's excursion, offered direction and backing, drawing upon their own encounters to assist her with exploring the unfamiliar waters of temperature control.

The journey turned into a pot for Lyra's self-improvement. Her experiences with legendary animals and the natural powers that watched the way tried the constraints of her command over intensity and cold. However, with every preliminary survive, Lyra's authority over temperature control extended. The Fire, reverberating with the agreeable exchange of components, directed her towards a significant comprehension of the enchanted that flowed through her veins.

In the magical mazes, where the reverberations of Aeliana's penance waited, Lyra went up against the shadows inside her own spirit. The Fire, sensitive to the immaculateness of her expectations, uncovered the weaknesses and fears that took steps to stifle the splendor of her true capacity. Lyra, sustained by the examples of the people who had preceded her, rose up out of the cauldron with newly discovered strength and an assurance to embrace the obligations that accompanied her novel gift.

As Lyra's process unfurled, the Flamebearers perceived the development of another time in Eldoria — a part where temperature control turned into a powerful power, a supplement to the fragile dance among light and shadow. The magical wards, once receptive to the energies of the Fire, reverberated with Lyra's capacities, extending their defensive reach to incorporate the subtleties of fire and ice. The domains, previously prospering affected by the Timeless Fire, answered the essential equilibrium that Lyra brought, making an embroidery of wizardry that rose above the limits of conventional magic.

The board of Flamebearers, directed by the Fire's immortal insight, embraced Lyra's job as a steward of temperature control. Together, they thought on the likely utilizations of her capacities and the effect they could have on Eldoria's sensitive equilibrium. The Fire, a quiet spectator and guide, offered bits of knowledge into the interconnectedness of enchantment, indicating the neglected profundities of temperature control that could shape the actual groundworks of the domain.

Individuals of Eldoria, seeing the development of Lyra and the unknown outskirts of temperature control, viewed her with a combination of stunningness and respect. Her story became entwined with the remarkable people of Aeliana and the Flamebearers, a demonstration of the steadily developing nature of sorcery in their reality. Pioneers, propelled by Lyra's excursion, left on new odysseys to investigate the secrets of temperature control, looking to outfit its power to improve their networks.

As Eldoria prospered under the amicable transaction of basic powers, murmurs of likely continuations and side projects reverberated through the captivated forests.

The Flamebearers, each with their one of a kind gifts and predeterminations, became heroes by their own doing, forming the fate of Eldoria in manners that rose above the impediments of the past. The Fire, timeless and all-knowing, indicated domains yet neglected, where new undertakings anticipated those ready to step the strange ways of wizardry.

Among the Flamebearers, a soothsayer named Kael arose with the capacity to control the texture of time itself. His excursion, set against the scenery of Eldoria's fleeting embroidery, unfurled as an adventure of transient control and the results that reverberated through the ages. The Fire, resounding with the complexities of time enchantment, alluded to the potential for worldly undertakings that could disentangle the secrets of the past and shape the predeterminations representing things to come.

As Kael dove into the intricacies of transient control, the domains shuddered with the repercussions of his decisions. The enchanted wards, woven into the texture of Eldoria by the Flamebearers, answered the subtleties of time wizardry, offering looks into substitute courses of events and expected prospects. The chamber, perceiving the meaning of Kael's excursion, wrestled with the sensitive equilibrium expected to employ the force of time without unwinding the actual groundworks of the real world.

Individuals of Eldoria, impacted by the interaction of temperature control and transient wizardry, saw the development of their reality into an embroidery of potential outcomes. The Fire, its gleam undiminished, kept on consuming at the core of Eldoria, projecting its brilliant light across the domains. Pioneers, drawn by the appeal of neglected outskirts, set out on ventures that rose above the limits of regular enchantment, trying to uncover the undiscovered likely that lay secret inside the texture of their reality.

As the Flamebearers and their exceptional capacities molded the fate of Eldoria, the murmurs of expected continuations and side projects turned into a tune that reverberated through the charmed forests. New heroes, each with their own gifts and predeterminations, arose to set out on ventures that extended the limits of wizardry and investigated domains yet unseen. The Fire, everlasting and omniscient, indicated the strings of predetermination that kept on winding through the embroidery of Eldoria, welcoming new ages to become stewards of the domain's wizardry and watchmen of its timeless fire.

In the supernatural domain of Eldoria, where the fragile dance of wizardry wove through the texture of the real world, another period unfolded with the development of temperature control — a strong power that tackled the transaction of intensity and cold to shape the actual embodiment of the world. The brilliant shine of the Everlasting Fire washed the land in an immortal hug, its brilliant light projecting energetic shades across the charming scenes.

At the core of this change stood Lyra, a Flamebearer whose natural capacity to control temperature set her on an odyssey to open the insider facts of this unfamiliar enchantment. Her process reflected the journey embraced by Aeliana, the incredible Flamebearer who had risen above human impediments to converge with the Everlasting Fire. Lyra's fate, nonetheless, unfurled along the neglected outskirts of natural equilibrium, where fire and ice entwined in an amicable dance.

The journey initiated with Lyra wandering into the captivated woodlands, where antiquated trees murmured stories of the secrets that looked for her. The Fire, sensitive to the virtue of her aims, reverberated with the agreeable exchange of intensity and cold that flowed through her veins. The supernatural wards, woven by the Flamebearers to defend Eldoria, detected the development of another power and answered with an unobtrusive change in their defensive energies.

As Lyra explored the slippery scenes and confronted the preliminaries intended to test her dominance over temperature control, she found the significant capability of her capacities. The Fire, a quiet aide and observer to her excursion, offered bits of knowledge into the fragile equilibrium expected to control basic powers without disturbing the harmony of Eldoria.

The experiences with legendary animals became valuable open doors for Lyra to refine her command over fire and ice. Within the sight of these creatures, epitomes of Eldoria's enchantment, she figured out how to explore the intricacies of temperature control with artfulness. The Flamebearers, perceiving the meaning of her excursion, offered direction drawn from their own encounters, manufacturing a brotherhood that rose above the limits of individual missions.

In the otherworldly mazes, where reverberations of Aeliana's penance waited, Lyra went up against the shadows inside her own spirit. The Fire, a wellspring of faithful brightening, uncovered the weaknesses and fears that took steps to douse the splendor of her true capacity. It turned into a snapshot of significant thoughtfulness, a defining moment where Lyra rose up out of the cauldron with freshly discovered flexibility and an assurance to embrace the obligations gave to her by the Fire.

The otherworldly wards, ever receptive to the subtleties of enchantment, adjusted to Lyra's presence. The defensive charms extended to consolidate the basic equilibrium she brought, offering extra layers of guard against likely dangers. Eldoria itself appeared to answer the development of temperature control, as the air sparkled with the interchange of fire and ice, making a supernatural atmosphere that enamored the faculties.

As Lyra's process unfurled, the board of Flamebearers met in the holy forests to consider on the meaning of her capacities.

The Fire, an immortal spectator and guide, offered bits of knowledge into the possible uses of temperature control and the effect it could have on Eldoria's sensitive equilibrium. The chamber, perceiving the development of another time, embraced

Lyra's job as a steward of this exceptional enchantment and promised to help her in the difficulties that lay ahead.

Individuals of Eldoria, observers to the extraordinary force of temperature control, viewed Lyra with a combination of stunningness and love. Her story became entwined with the remarkable people of Aeliana and the Flamebearers, a demonstration of the steadily developing nature of wizardry in their reality. Travelers, motivated by Lyra's excursion, left on new odysseys to investigate the secrets of temperature control, trying to outfit its power to improve their networks.

The captivated urban communities of Eldoria, enhanced with towers that came to towards the sky, became center points of mystical advancement. Researchers and sorcerers dove into the complexities of temperature control, trying to open its likely applications in horticulture, development, and, surprisingly, the improvement of regular day to day existence. The business sectors hummed with the trading of information, and the actual air turned into a material for experts to paint energetic presentations of fire and ice.

However, as Eldoria prospered affected by the agreeable exchange of natural powers, murmurs of possible continuations and side projects reverberated through the charmed forests. The Flamebearers, each with their exceptional gifts and fates, became heroes by their own doing, molding the eventual fate of Eldoria in manners that rose above the constraints of the past. The Fire, everlasting and all-knowing, indicated domains yet neglected, where new undertakings anticipated those able to step the unknown ways of sorcery.

Among the Flamebearers, a soothsayer named Kael arose with the capacity to control the texture of time itself. His excursion, set against the background of Eldoria's fleeting embroidery, unfurled as an adventure of worldly control and the results that reverberated through the ages. The Fire, reverberating with the complexities of time enchantment, indicated the potential for fleeting undertakings that could disentangle the secrets of the past and shape the fates representing things to come.

As Kael dug into the intricacies of fleeting control, the domains trembled with the repercussions of his decisions. The otherworldly wards, woven into the texture of Eldoria by the Flamebearers, answered the subtleties of time enchantment, offering looks into substitute courses of events and possible fates. The chamber, perceiving the meaning of Kael's excursion, wrestled with the sensitive equilibrium expected to use the force of time without unwinding the actual underpinnings of the real world.

Individuals of Eldoria, impacted by the transaction of temperature control and fleeting wizardry, saw the development of their reality into an embroidery of conceivable outcomes. The Fire, its sparkle undiminished, kept on consuming at the core of Eldoria, projecting its brilliant light across the domains. Travelers, drawn by the charm of neglected wildernesses, left on ventures that rose above the limits of traditional wizardry, trying to uncover the undiscovered possible that lay secret inside the texture of their reality.

As the Flamebearers and their remarkable capacities formed the predetermination of Eldoria, the murmurs of possible continuations and side projects turned into an ensemble that reverberated through the captivated forests. New heroes, each with their own gifts and predeterminations, arose to leave on ventures that extended the limits of enchantment and investigated domains yet unseen. The Fire, everlasting and omniscient, alluded to the strings of predetermination that kept on winding through the embroidered artwork of Eldoria, welcoming new ages to become stewards of the domain's enchantment and gatekeepers of its timeless fire.

Conclusion

"Mercury Increasing: Stories of Temperature" is an arresting investigation of the diverse idea of temperature and its significant effect on different parts of our lives. All through this excursion, we have dug into the multifaceted dance among intensity and cold, analyzing their effect on the climate, human wellbeing, innovation, and cultural elements. As we wind down this exhaustive investigation, it becomes obvious that temperature isn't simply a mathematical worth on a thermometer; a power shapes our reality in manners both seen and concealed.

At its center, "Mercury Rising" has unwound the many-sided connection among temperature and the regular world. From the climbing temperatures related with environmental change to the fragile equilibrium expected for biological systems to flourish, we have seen the significant results of temperature variances. The stories of liquefying glacial masses, moving environment zones, and the situation of jeopardized species act as obvious tokens of the dire requirement for worldwide participation to address the difficulties presented by a warming planet.

The human component of temperature shapes a critical string in the story, featuring the fragile harmony between our bodies and the climate. The effects of outrageous heatwaves on general wellbeing, farming, and weak networks have been revealed. We have investigated the versatile techniques utilized by people and social orders to adapt to the evolving environment, perceiving that strength and advancement are essential parts of our aggregate reaction to climbing temperatures.

Innovation, an integral asset in our mission to comprehend and control temperature, has been a focal person in this adventure. From the origin of old cooling strategies to the state of the art developments in environment science and warm designing, we have seen the advancement of human creativity. The stories of mechanical leap forwards, for example, the improvement of reasonable energy sources and the journey for proficient cooling frameworks, highlight our steady quest for answers for relieve the unfavorable impacts of temperature limits.

As we cross the scenes of "Mercury Rising," the story strings combine on the vital job of temperature in significantly shaping social orders. From social practices and ceremonies attached in climatic circumstances to the international ramifications of asset shortage, the narratives unfurl a complicated embroidery of human communications with temperature. The acknowledgment that temperature impacts our prompt environmental factors as well as the complex snare of worldwide connections accentuates the interconnectedness of our reality.

In the domain of wellbeing, the effect of temperature on the prosperity of people has been carefully analyzed. From the physiological reactions of the human body to warm pressure to the unpretentious impact of temperature on emotional well-being, we have gotten a handle on the significant ramifications for medical services frameworks and public strategies. The stories of irresistible infections impacted by temperature vacillations act as strong tokens of the mind boggling interaction between the tiny world and the more extensive natural setting.

In the unpredictable dance of temperature, farming arises as a vital participant, its destiny laced with the impulses of environment. The accounts of changing developing seasons, adjusted precipitation designs, and the apparition of food weakness highlight the weakness of our horticultural frameworks to temperature limits. However, in the midst of these difficulties, accounts of versatile cultivating rehearses and the mission for environment tough harvests offer looks at trust and flexibility despite an evolving environment.

The general subject of "Mercury Rising" isn't one of gloom, yet rather a source of inspiration. The stories of temperature, both warm and cold, coax us to face the truth of a changing environment and to perceive the significant obligation we bear as stewards of this planet. The criticalness of relieving environmental change, adjusting to its unavoidable effects, and cultivating a practical conjunction with temperature limits penetrates the story, provoking us to rise above limits and manufacture an aggregate way ahead.

As we ponder the different stories woven into the texture of "Mercury Rising," the general story is clear: temperature is a powerful power that shapes the past, present, and fate of our reality. Whether appeared in the constant climb of worldwide temperatures or the complex expressive dance of atomic movement, temperature is a ubiquitous component that requests our consideration and regard. The stories of temperature typify the logical standards administering warm elements as well as the profoundly human accounts of variation, advancement, and versatility notwithstanding ecological difficulties.

All in all, "Mercury Climbing: Stories of Temperature" rises above the limits of a traditional investigation of logical standards. An embroidery of stories welcomes perusers to examine the significant ramifications of temperature on our planet and our lives. From the minute domain of sub-atomic movement to the huge territories of the World's environment frameworks, the tales told inside these pages wind around together a complete comprehension of temperature and its expansive outcomes.

As we explore the intricacies of a reality where mercury rises and falls, the test before us is clear. We should recognize the truth of an evolving environment, stand up to the weaknesses uncovered by temperature limits, and on the whole take a stab at reasonable arrangements. The tales inside "Mercury Rising" act as signals of information, directing us towards a future where our relationship with temperature is one of equilibrium, stewardship, and shared liability.

Eventually, "Mercury Rising" is in excess of an assortment of stories — it is a source of inspiration. The stories of temperature, with every one of their subtleties and intricacies, coax us to draw in with the difficulties and potential open doors introduced by a world in motion. As the mercury proceeds with its dance, let us notice the examples gained from these stories and work towards a future where the increasing temperatures of our planet are tempered by the insight, development, and aggregate purpose of mankind.

www.ingramcontent.com/pod-product-compliance
Lightning Source LLC
LaVergne TN
LVHW010339200726
843507LV00010B/1557